T0320789

# Introductory Computational Physics

Computers are one of the most important tools available to physicists, whether for calculating and displaying results, simulating experiments, or solving complex systems of equations.

Introducing students to computational physics, this textbook shows how to use computers to solve mathematical problems in physics and teaches students about choosing different numerical approaches. It also introduces students to many of the programs and packages available. The book relies solely on free software: the operating system chosen is Linux, which comes with an excellent C++ compiler, and the graphical interface is the ROOT package available for free from CERN.

This up-to-date, broad scope textbook is suitable for undergraduates starting on computational physics courses. It includes exercises and many examples of programs. Online resources at www.cambridge.org/9780521828627 feature additional reference information, solutions, and updates on new techniques, software and hardware used in physics.

ANDI KLEIN is a Technical Staff member at Los Alamos National Laboratory, New Mexico. He gained his Ph.D. from the University of Basel, Switzerland. He held the position of Professor of Physics at Old Dominion University, Virginia, from 1990 to 2002, where he taught courses in computational physics.

ALEXANDER GODUNOV is Assistant Professor at the Department of Physics, Old Dominion University, Virginia. He gained his Ph.D. from Moscow State University, Russia and has held research positions at Tulane University, Louisiana, and visiting positions at research centers in France and Russia.

# Introductory Computational Physics

**Andi Klein and Alexander Godunov**

Los Alamos National Laboratory
and Old Dominion University

CAMBRIDGE
UNIVERSITY PRESS

# CAMBRIDGE
### UNIVERSITY PRESS

University Printing House, Cambridge CB2 8BS, United Kingdom

Published in the United States of America by Cambridge University Press, New York

Cambridge University Press is part of the University of Cambridge.

It furthers the University's mission by disseminating knowledge in the pursuit of education, learning and research at the highest international levels of excellence.

www.cambridge.org
Information on this title: www.cambridge.org/9780521828628

© Cambridge University Press 2006

First published 2006

A catalogue record for this publication is available from the British Library

ISBN  978-0-521-82862-8  Hardback
ISBN  978-0-521-53562-5  Paperback

# Contents

# Preface

Computers are one of the most important tools in any field of science and especially in physics. A student in an undergraduate lab will appreciate the help of a computer in calculating a result from a series of measurements. The more advanced researcher will use them for tasks like simulating an experiment, or solving complex systems of equations. Physics is deeply connected to mathematics and requires a lot of calculational skills. If one is only interested in a conceptual understanding of the field, or an estimate of the outcome of an experiment, simple calculus will probably suffice. We can solve the problem of a cannon ball without air resistance or Coriolis force with very elementary math, but once we include these effects, the solution becomes quite a bit more complicated. Physics, being an experimental science, also requires that the measured results are statistically significant, meaning we have to repeat an experiment several times, necessitating the same calculation over and over again and comparing the results. This then leads to the question of how to present your results. It is much easier to determine the compatibility of data points from a graph, rather than to try to compare say 1000 numbers with each other and determine whether there is a significant deviation. From this it is clear that the computer should not only "crunch numbers," but should also be able to display the results graphically.

Computers have been used in physics research for many years and there is a plethora of programs and packages on the Web which can be used to solve different problems. In this book we are trying to use as many of these available solutions as possible and not reinvent the wheel. Some of these packages have been written in FORTRAN, and in Appendix C you will find a description of how to call a FORTRAN subroutine from a C++ program. As we stated above, physics relies heavily on graphical representations. Usually, the scientist would save the results from some calculations into a file, which then can be read and used for display by a graphics package like **gnuplot** or a spreadsheet program with graphics capability. We have decided to pursue a different path, namely using the **ROOT** package [1] developed at the high energy physics lab CERN in Switzerland. ROOT, being an object oriented C++ package, not only provides a lot of physics and math C++-classes but also has an excellent graphics environment, which lets you create publication quality graphs and plots. This package is constantly being developed and new features and classes are being added. There is an excellent user's guide, which can be found on the ROOT website in different formats. In order to get started quickly we have given a short introduction in Appendix A.

# Chapter 1
# Introduction

## 1.1 The need for computers in science

Over the last few decades, computers have become part of everyday life. Once the domain of science and business, today almost every home has a personal computer (PC), and children grow up learning expressions like "hardware," "software," and "IRQ." However, teaching computational techniques to undergraduates is just starting to become part of the science curriculum. Computational skills are essential to prepare students both for graduate school and for today's work environment.

Physics is a corner-stone of every technological field. When you have a solid understanding of physics, and the computational know-how to calculate solutions to complex problems, success is sure to follow you in the high-tech environment of the twenty-first century.

## 1.2 What is computational physics?

Computational physics provides a means to solve complex numerical problems. In itself it will not give any insight into a problem (after all, a computer is only as intelligent as its user), but it will enable you to attack problems which otherwise might not be solvable. Recall your first physics course. A typical introductory physics problem is to calculate the motion of a cannon ball in two dimensions. This problem is always treated without air resistance. One of the difficulties of physics is that the moment one goes away from such an idealized system, the task rapidly becomes rather complicated. If we want to calculate the solution with real-world elements (e.g., drag), things become rather difficult. A way out of this mess is to use the methods of computational physics to solve this linear differential equation.

One important aspect of computational physics is modeling large complex systems. For example, if you are a stock broker, how will you predict stock market performance? Or if you are a meteorologist, how would you try to predict changes in climate? You would solve these problems by employing Monte Carlo techniques. This technique is simply impossible without computers and, as just noted, has applications which reach far beyond physics.

Another class of physics problems are phenomena which are represented by nonlinear differential equations, like the chaotic pendulum. Again, computational physics and its numerical methods are a perfect tool to study such systems. If these systems were purely confined to physics, one might argue that this does not deserve an extended treatment in an undergraduate course. However, there is an increasing list of fields which use these equations; for example, meteorology, epidemiology, neurology and astronomy to name just a few.

An advantage of computational physics is that one can start with a simple problem which is easily solvable analytically. The analytical solution illustrates the underlying physics and allows one the possibility to compare the computer program with the analytical solution. Once a program has been written which can handle the case with the typical physicist's approximation, then you add more and more complex real-world factors.

With this short introduction, we hope that we have sparked your interest in learning computational physics. Before we get to the heart of it, however, we want to tell you what computer operating system and language we will be using.

## 1.3 Linux and C++

### Linux

You may be accustomed to the Microsoft Windows or Apple MAC operating systems. In science and in companies with large computing needs, however, UNIX is the most widely used operating system platform. Linux is a UNIX-type operating system originally developed by Linus Torwald which runs on PCs. Today hundreds of people around the world continue to work on this system and either provide software updates or write new software. We use Linux as the operating system of choice for this text book because:

- Linux is widely available at no cost;
- Linux runs on almost all available computers;
- it has long-term stability not achieved by any other PC operating system;
- Linux distributions include a lot of free software, i.e., PASCAL, FOR-TRAN, C, C++.

In today's trend to use networked clusters of workstations for large computational tasks, knowledge of UNIX/Linux will provide you with an additional, highly marketable skill.

## C++

In science, historically the most widely used programming language was FORTRAN, a fact reflected in all the mathematical and statistical libraries still in use the world over (e.g., SLATEC, LAPACK, CERNLIB). One disadvantage of FORTRAN has always been that it was strongly decoupled from the hardware. If you wanted to write a program which would interact directly with one of the peripherals, you would have to write code in assembly language. This meant that not only had you to learn a new language, but your program was now really platform dependent.

With the emergence in the late 1970s of C [2] and UNIX, which is written in C, all of a sudden a high level language was available which could do both. C allowed you to write scientific programs and hardware drivers at the same time, without having to use low level processor dependent languages. In the mid 1980s Stroustrup [3] invented C++, which extended C's capabilities immensely. Today C and C++ are the most widely used high level languages.

Having "grown up" in a FORTRAN environment ourselves, we still consider this to be the best language for numerical tasks (we can hear a collective groan in the C/C++ community). Despite this, we decided to "bite the bullet" and switch to C++ for the course work.

The GNU C/C++ compiler is an excellent tool and quite versatile. Compared to the **Windows** C++ compilers (e.g., Visual C++ [Microsoft] or Borland C/C++), the user interface is primitive. While the Windows compiler-packages have an extensive graphical user interface (GUI) for editing and compiling, the GNU compiler still requires you first to use a text editor and then to collect all the necessary routines to compile and link. One "disadvantage" to the Windows compiler packages is that many of them automatically perform a number of tasks necessary to building a program. You might be wondering how that could be a disadvantage. We have noticed that when students have used such packages, they often have a poor understanding of concepts like linking, debuggers, and so on. In addition, if a student switches from one Windows compiler package to another, s/he must learn a new environment. Therefore, this text will use/refer to the GNU C/C++ compiler; however the programs can be easily transported to the afore mentioned proprietary systems.

Having extolled the virtues of C/C++, we must mention here that some of
the sample programs in this book reflect our roots in FORTRAN. There are
many functions and subroutines available for scientific tasks which have been
written in FORTRAN and have been extensively tested and used. It would
be foolish to ignore these programs or attempt to rewrite them in C/C++. It
is much less time consuming to call these libraries from C++ programs than
it is to write your own version. In Appendix C we describe how FORTRAN
libraries and subroutines can be called from C++.

# Chapter 2
# Basics

Before we start we need to introduce a few concepts of computers and the interaction between you, the user, and the machine. This will help you decide when to write a program for solving a physics or science problem and when it is much easier or faster to use a piece of paper and a pocket calculator. In thinking about computers, remember there is a distinction between hardware and software. Software is divided into the operating system and your particular application, like a spreadsheet, word-processor or a high level language. In this book we will spend most of the time in dealing with issues relevant to physics and the algorithms used to solve problems. However, in order to make this as productive as possible, we will start off with a short description of the hardware and then some discussion of the operating system.

## 2.1 Basic computer hardware

Apart from huge parallel supercomputers, all workstations you can buy today are organized in a similar way (Figure 2.1).

The heart of the computer is the **CPU** (Central Processing Unit) controlling everything in your workstation. Any disk I/O (Input/Output) or computational task is handled by the CPU. The speed at which this chip can execute an instruction is measured in Hz (cycles per second) at several GHz. The CPU needs places to store data and instructions. There are typically four levels of memory available: level I cache, level II cache, **RAM** (Random Access Memory) and **swap space**, the last being on the hard disk. The main distinction between the different memory types is the speed.

The cache memory is located on the CPU chip.

This CPU chip has two small memory areas (one for data and one for instructions), called the **level I caches** (see below for further discussion),

5

**Figure 2.1** Schematic
layout of a workstation.

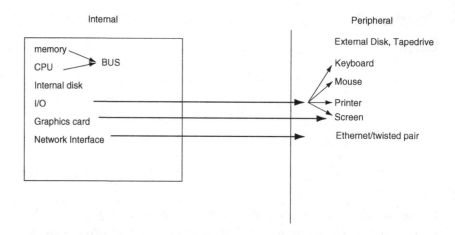

which are accessed at the full processor speed. The second cache, level II, acts as a fast storage for code or variables needed by the code. However, if the program is too large to fit into the second cache, the CPU will put some of the code into the RAM. The communication between the CPU and the RAM is handled by the **Bus**, which runs at a lower speed than the CPU clock. It is immediately clear that this poses a first bottleneck compared to the speed the CPU would be able to handle. As we will discuss later, careful programming can help in speeding up the execution of a program by reducing the number of times the CPU has to read and write to RAM. If this additional memory is too small, a much more severe restriction will come into play, namely **virtual memory** or **swap space**. Virtual memory is an area on the disk where the CPU can temporarily store code which does not fit into the main memory, calling it in when it is needed. However, the communication speed between the CPU and the virtual memory is now given by the speed at which a disk can do I/O operations.

The internal disk in Figure 2.1 is used for storing the operating system and any application or code you want to keep. The disk size is measured in gigabytes (**GB**) and 18–20 GB disks are standard today for a workstation. Disk prices are getting cheaper all the time thus reducing the need for code which is optimized for size. However the danger also is that people clutter their hard disk and never clean it up.

Another important part of your system is the Input/Output (**I/O**) system, which handles all the physical interaction between you and the computer as well as the communication between the computer and the external peripherals (e.g., printers). The I/O system responds to your keyboard or mouse but will also handle print requests and send them to the printer, or communicate with an external or internal tape or hard drive.

The last piece of hardware we want to describe briefly is the network interface card, which establishes communication between different computers over a network connection. Many home users connect to other computers through a modem, which runs over telephone lines. Recently cable companies have started to offer the use of their cable lines for network traffic, allowing the use of faster cable modems. The telephone modem is currently limited to a maximum speed of 56 kB/s, which will be even slower if the line has a lot of interferences. In our environment we are using an Ethernet network, which runs at 100 MB/s.

## 2.2 Software

### Operating system

In getting your box to do something useful, you need a way to communicate with your CPU. The software responsible for doing this is the **operating system** or OS. The operating system serves as the interface between the CPU and the peripherals. It also lets you interact with the system. It used to be that different OSs were tied to different hardware, for example VMS was Digital Equipment's (now Hewlett-Packard) operating system for their VAX computers, Apple's OS was running on Motorola chips and Windows 3.1 or DOS was only running on Intel chips. This of course led to software designers concentrating on specific platforms, therefore seriously hampering the distribution of code to different machines. This has changed and today we have several OSs which can run on different platforms.

One of the first operating systems to address this problem was UNIX, developed at the AT&T Labs. Still a proprietary system, it at least enabled different computer manufacturers to get a variant of UNIX running on their machines, thus making people more independent in their choices of computers. Various blends of UNIX appeared as shown in the following table:

| | |
|---|---|
| Ultrix | Digital Equipment Corporation (now HP) |
| HP True64 Unix (formerly OSF) | HP |
| Solaris/SunOS | SUN |
| AIX | IBM |
| HP-Unix | Hewlett-Packard |

These systems are still proprietary and you cannot run HP-Unix on a Sun machine and vice versa, even though for you as a user they look very similar. In the 1980s Linus Torwald started a project for his own amusement called Linux, which was a completely free UNIX system covered under the GNU license. Since then many people have contributed to Linux and it is now a mature and stable system. In the scientific community Linux is the fastest growing operating system, due to its stability and low cost and its ability to run on almost all computer platforms. You can either download Linux over the internet from one of the sites listed in the appendix, or you can buy one of the variants from vendors like Red Hat or SuSE. The differences in these distributions are in ease of installation, graphical interfaces and support for different peripherals. However the **kernel**, the heart of the operating system, is the same for all.

Unlike other operating systems, when you get Linux, you also get the complete source code. Usually, you do not change anything in the code for Linux, unless either you are very knowledgeable (but read the license information first) or you want to get into trouble really fast.

Two other advantages of Linux are that there is a lot of free application software and it is a very stable system. The machines in our cluster run for months without crashing or needing to be rebooted.

## Applications and languages

This is the part the average user is most familiar with. Most users buy applications out of the box, consisting of business software like spreadsheets, word processors and databases. For us the most important issue is the programming language. These are the languages you can use to instruct your computer to do certain tasks and are usually referred to as high level languages in contrast to assembly language.

We usually distinguish high level languages in the following way: interpreted languages like **Basic**, **Perl**, **awk** and compiled ones like **FORTRAN**, **FORTRAN90**, **C** and **C++**. The distinction, however, is not clear cut; there are C-interpreters and compiled versions of Perl. The interpreted languages execute every line the moment it is terminated by a carriage return and then wait for the next line. In a way this is similar to your pocket calculator, where you do one operation after the next. This is a very handy way of doing some calculations but it will pose some serious restrictions, especially when you try to solve more complex problems or you want to use libraries or functions which have been previously written. Another disadvantage is the slow running of the program and the lack of optimization.

In a compiled language you first write your complete code (hopefully without error), gather all the necessary functions (which could be in libraries) and then have the computer translate your entire program into machine language. Today's compilers will not only translate your code but will also try to optimize the program for speed or memory usage. Another advantage of the compiler is that it can check for consistency throughout the program, and try to catch some errors you introduced. This is similar to your sophisticated word processor, which can catch spelling and even some grammar mistakes. However, as the spell checker cannot distinguish between two or too (both would be fine) or check whether what you have written makes sense, so the compiler will not be able to ensure consistency in your program. We will discuss these issues further below, when we give our guidelines for good programming practice.

After you have run your different routines through the compiler, the last step is the linker or loader. This step will tie together all the different parts, reserve space in memory for variables, and bind any needed library to your program. On Linux the last step is usually executed automatically when you invoke the compiler. The languages most used in scientific computing (especially in physics) are FORTRAN and C/C++. Traditionally FORTRAN was the language of choice, and still today there is a wealth of programs readily available in FORTRAN libraries (e.g. **CERN library**, **SLATEC**, **LAPACK**). During the last decade, C/C++ has become more and more important in physics, so that this book focuses on C++ (sigh!) and moves away from FORTRAN. We are still convinced that FORTRAN is the better language for physics, but in order to get the newer FORTRAN90/95 compiler, one has to buy a commercial package, while the C and C++ compilers are available at no cost for Linux.

## 2.3 How does it work?

In Figure 2.2 we have outlined how the different layers on a UNIX workstation can be grouped logically. The innermost part is the kernel, which controls the hardware, where the services are the part of the system which interacts directly with the kernel. Assembly language code will interact with this system level. The utilities layer contains programs like rm (remove) or cp (copy) and the compilers. The user interface and program development

**Figure 2.2** Schematic
layout of a workstation.

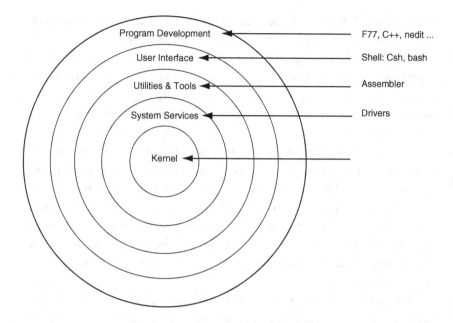

**Figure 2.2** Schematic layout of a workstation.

areas are where you will be working most. You can choose the particular shell you prefer, which then will be your interface to the lower levels. Even though you will be in an X-Window environment, you still have to use command line input and write scripts, which will automate your tasks. This is done in your chosen shell, and some of the commands will be different for different shells. In the outermost shell you will have your applications, like compiled programs.

# Chapter 3
# Short introduction to Linux

Unix: The world's first computer virus
*from The UNIX-Haters Handbook [4]*

## 3.1 Getting started and logging in

We will try to jump-start you into the Linux environment. The first thing you have to do is log into the system. Since Linux is a real multi-user system, the interaction between you and the computer might be different than what you are used to from a Microsoft or Macintosh environment. You could be either at the computer console or at a terminal, which is connected via a network to the computer. In either way, you will see a Windows-like screen which will display a login screen, asking you for the username and the password. Assuming that your system manager has set you up already, you will type in both, and as long as you did not mistype anything you should now be in the computer. In case you made a mistake in typing in either of the two items, the computer will not let you in. (Note that Linux is case sensitive, so **Emma** is not the same as **emma**.) Depending on the setup of your computer, you will now be faced with a graphical user interface (**GUI**), the most common of these being either **KDE** or **Gnome**. Click on the small icon which resembles a terminal. This will bring a new window, which lets you type in commands, somewhat like the **command** icon in DOS. If this is the first time you have logged into this account, you should change your password, especially if your system administrator has assigned one to you. The normal Linux command for changing the password is **passwd**, which you will have to type in. In our PC-farm environment we use the Network Information System, formerly known as the YP system, which has the password file for the whole cluster centralized. In this case you have to type in **yppasswd**, then answer the questions for the old password (so nobody unauthorized can change yours) and give the new password twice:

```
passwd
Old Password:
New Password:
New Password:
```

Make sure that your password is eight characters long and use special, lower and upper case characters. As an example of a bad password **emma**, a name comes to mind. However you can make this into a good password by using **$e5Mm%a!**.

## 3.2 Getting help

Now that you are set up you can start using Linux. The first thing you will probably do is cry for help. Linux, as every good UNIX system, is not very good in responding to your needs for help, and provides you with a completely archaic system to learn about specific commands. (If you are getting really disappointed with your lack of progress, log out for a while, get the *UNIX-Haters Handbook* [4], and read some professionals' opinions and frustations about UNIX.) In Linux, to get help you use the **man** command, which will let you look up the specifics of a command. The man stands for manual and is organized in different sections. However, the man command has its own page, so the first thing you want to do is

**man man**

which will explain how to use the man system and what kind of arguments and options the man command takes.

Most of the commands in Linux have options which change the behavior of the command or give you additional information. One important command is the **man -k** *blabla*, which will look up in the manual database anything that matches blabla. So in case you cannot remember the exact word you can try to find it with the **man -k** command. But let us warn you, some of the commands have very unintuitive names like **awk** or **cat**. Another problem with UNIX is that some of the commands have their own man pages like **wc, ls** or **awk**. To add to the confusion, UNIX systems let you choose your preferred command language, called a shell. **jobs** or **alias** belong to a particular shell and you have to read the complete man page for the particular shell you are using.

## 3.3 The filesystem, or where is everything?

In the following we will outline the typical Linux file system and how it is organized. This should make it easier for you to understand the operating system and learn how to find programs and resources on your computer.

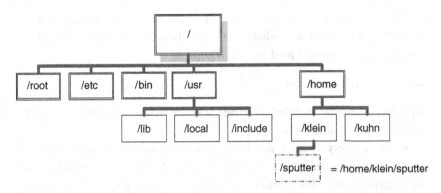

**Figure 3.1** The Linux file system.

As you can see from Figure 3.1, the top directory is called /. Everything is referenced in respect to this directory or folder. The location which you will be most concerned with is the /home/*your username*, which will be your home directory. You see a /home/klein, which would be the home directory of Andi Klein. The home directory always has the same name as you have chosen for your username.

There are a few more directories worth mentioning, the first being /root. This is the home directory of your system administrator, who in Linux is called "root." Root has special privileges such as the ability to shut down the machine, delete files and create new user accounts. However, because he or she has such strong privileges, they can also do devastating things to the system, like accidentally removing important system files and making the system inoperable. Every system adminstrator's worst nightmare is that some malicious person can get access to the system with root privileges and erase the system.

The next directories we want to mention are the /usr/lib and /usr/include. Here you will find libraries and include files for various programs which you might want to use. You can either address these directories directly by specifying their complete path every time you need something from there or you can have the path included in your .login command file, which is explained below.

## 3.4 Moving around in your system

When you log into your system, you usually land by default in your own so-called home directory. In our case this would be /home/klein. This is where you work and create your programs. By executing **pwd**, you can check the full name of your home directory. **pwd** is a very helpful command, because it will always let you know where you currently are in the file system. Here is the output of a typical **pwd** on our system:

```
/home/klein/comp_phys
```

telling us that we are in directory **comp_phys**, which is a subdirectory of the login directory **/home/klein**.

This is not very helpful yet, because you do not know how to move around. Be patient, this will come in a short while. First, you should create some subdirectories or folders. Tis will help you to keep your working area organized and lets you find programs or other important stuff again later. To create a directory type **mkdir** *directory name*. Typing **cd** *dir*, meaning change directory to *dir*, you can go into this newly created subdirectory.

## 3.5 Listing your directory

Now that you know how to move around and create directories, it is time to discuss how you list the content. This is done with **ls** which will give you a bare bones listing of almost all the files you have in your directory (Figure 3.2).

In the second part of Figure 3.2 we have chosen the option **-al** which gives all the so-called **hidden** files, files which start with a ., and also lists the directories. In addition it gives the ownership and the permissions **rwx** for each file.

A permission of **w** means that you can write to this file, or even more importantly, delete this file. So make sure that any file has only write permission in the first column, namely the permission for the owner. As we just mentioned, you can delete your files, and the command to do this is **rm**

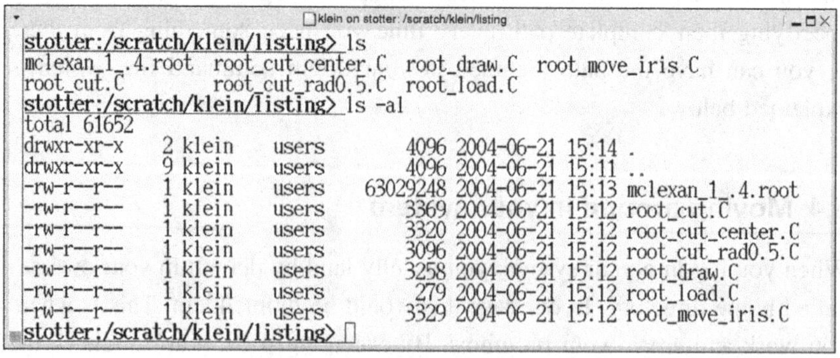

```
klein on stotter: /scratch/klein/listing
stotter:/scratch/klein/listing> ls
mclexan_1_.4.root  root_cut_center.C  root_draw.C  root_move_iris.C
root_cut.C         root_cut_rad0.5.C  root_load.C
stotter:/scratch/klein/listing> ls -al
total 61652
drwxr-xr-x    2 klein    users        4096 2004-06-21 15:14 .
drwxr-xr-x    9 klein    users        4096 2004-06-21 15:11 ..
-rw-r--r--    1 klein    users    63029248 2004-06-21 15:13 mclexan_1_.4.root
-rw-r--r--    1 klein    users        3369 2004-06-21 15:12 root_cut.C
-rw-r--r--    1 klein    users        3320 2004-06-21 15:12 root_cut_center.C
-rw-r--r--    1 klein    users        3096 2004-06-21 15:12 root_cut_rad0.5.C
-rw-r--r--    1 klein    users        3258 2004-06-21 15:12 root_draw.C
-rw-r--r--    1 klein    users         279 2004-06-21 15:12 root_load.C
-rw-r--r--    1 klein    users        3329 2004-06-21 15:12 root_move_iris.C
stotter:/scratch/klein/listing> 
```

**Figure 3.2** Listing of directory from **ls** and **ls -al**.

*filename*. This is a command you should use carefully, because once you have removed a file, it is gone. A safe way to prevent disaster is to alias rm to **rm -i**, which will ask you for confirmation before a file goes into the black hole. Here is a list of the commands covered so far:

| | |
|---|---|
| ls | list files and directories |
| ls -al | list files with . and permissions |
| cd *directory* | go to *directory* |
| cd | go to home (login) directory |
| cd ~ | go to home (login) directory |
| cd .. | change to parent directory of the current one |
| mkdir *name* | create directory with name *name* |
| rm *file* | delete file *file* (be careful) |
| pwd | where am I? |

## 3.6 Creating your own files

In order to create a file you can either type **touch** *filename*, which creates an empty file, or you can use an editor. Our preference is **nedit**, which, if the file does not exist, will ask you first if you want to create it. nedit is an X-Windows editor, i.e., you need to be on a system with X running. Once nedit is running you can click on the different menus on the window (Figure 3.3).

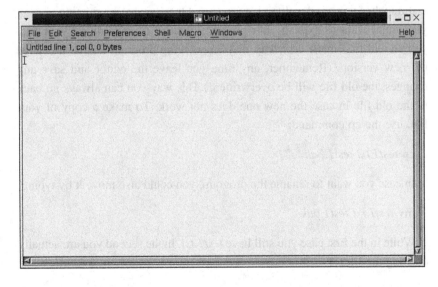

**Figure 3.3** The **nedit** editor.

If you are only on a regular terminal, the editor to use is **vi**. For this introduction, the description of vi would be too long. If you want more information on the vi command, get a book about UNIX; all of them have a description of the different commands. There is also **emacs**, which can run on either a terminal or an X-Window system. Although this is an excellent editor, it is fairly complicated to learn. Having created a file and saved it, say as *test1.txt*, you can then continue to work with it. One way to look at the content of the file is to open it again with your editor, or by using the **less** or **cat** commands. For example:

> **less** *test1.txt*

will list the content of the file one page at a time, and for longer files you can use the up and down arrows to scroll forward or backward in your text. **cat** will immediately dump everything on your screen, which can be rather annoying if the file consists of several hundred lines. A way out of this conundrum is to use the pipeline mechanism of UNIX, which lets you "pipe" one command into the next one. You invoke the pipe with the vertical bar I, which separates the commands. For those of you who remember the DOS days, that should be familiar. In our case, we want to pipe the **cat** command into the **more** command which stops after every page:

> **cat** *test1.txt* I **more**

Let us assume that you want to rename this file. This is an important safety feature when you are developing a program. Once you have a running version you might want to expand your program. Instead of just editing your file, you should first make a backup of your working program and then change the new version. (Remember, any time you leave the editor and save any changes, the old file will be overwritten!) This way you can always go back to the old file in case the new one does not work. To make a copy of your file, use the **cp** command:

> **cp** *test1.txt test1.bak*

In case you want to rename the program, you could also move it by typing:

> **mv** *test1.txt test1.bak*

While in the first case you still have *test1.txt*, in the second you are actually deleting the file.

| cp *old new* | copies file *old* to *new* |
| mv *old new* | renames file *old* to *new* |
| rm *file* | deletes file |
| rm -i *file* | deletes file, but asks for confirmation first |
| rmdir *directory* | deletes a directory |
| cat *file* | lists file content |
| cat *file* \| more | lists file content one page at a time |
| less *file* | lists file content with cursor control |

## 3.7 Doing some work

In order to tailor the system to your own preferences, first set up your environment. In the c-shell you have two files which will control your session. The first one is the **.login** (hidden) file which is automatically executed every time you log in. Here you can set or redefine environment variables or give the search path. The environment variables are independent of the shell you are using; i.e., if you need to execute a different shell these variables will stay. In the following we give a sample .login file, which you can use as a skeleton and add or change according to your own needs:

```
###########Begin .login file
#sample .login file
# first set up what is in the path, i.e. where the system looks
  for programs
if ($?path) then
  set path=($HOME/bin $HOME/util /usr/bin/X11 /usr/local/bin
  $path .)
else
  set path=($HOME/bin $HOME/util /usr/bin /usr/bin/X11
  /usr/local/bin /usr/lib .)
endif

setenv MANPATH /usr/local/pgqsl/man:/usr/share/man:

# here we set the prompt (works only with tcsh)
set prompt="%B%U%m:%~>"
#
#This creates a buffer of 40 previous commands
# which you can recall
set history=40
```

```
# the alias section

alias netscape '/opt/netscape/netscape &'
# the nexto command will only list directories
alias lsd 'ls\index{ls} -l | grep drwx'
# here we set some environment variables

set correct=cmd # spell checking for command line Tcsh
set notify #tells\index{ls} me about finished background jobs
#environment variables
setenv EDITOR /usr/bin/X11/nedit

#Setting up the root system
setenv ROOTSYS $SOFT/root_new
set path=($path $ROOTSYS/bin)
setenv LD_LIBRARY_PATH $ROOTSYS/lib
############# End of .login\index{.login} file
```

During this course, you will be writing your own programs. Linux has an excellent C/C++ compiler: **gcc** or **g++**. You can also use the FORTRAN77 compiler, which you invoke with **g77**. Here the help system is different; instead of man C you would type **info gcc**, or if **Tk/Tcl** is installed you can use the graphical interface **tkinfo**. The info package lets you navigate around the pages like you would on a Web page. There are several options for the compiler worth mentioning here to get you started, which we show in the following table.

| Option | Effect |
|---|---|
| -c | compile only, create an object file .o |
| -o filename | compile and link, name the executable filename |
| -Ox | optimization level x, where x can be 0 or 1,2,3 |
| -L | for linker, gives the library path |
| -l | for linker, gives the library name |
| -I | gives path for include files |
| -g | produces a debugging version |

If you need to debug your program, you should turn off optimization **g0** and set the flag **-g** in your compilation. Now you can use the gnu debugger **gdb** or again for an X-window system **ddd** (Figure 3.4).

**Figure 3.4** The **DDD** debugger window.

## 3.8 Good programming

One of the most common mistakes we see with our students is their rush to go and start "hacking in a program." You can save yourself a lot of headache and time if you think before you type. Here are a few guidelines to help you in being successful with your computer.

1. Do you really need a program for your task? This might sound silly, but consider the following problem:

$$\sum_{k=1}^{\infty} \frac{1}{2^k k} = ?$$

Now you will immediately say: this is a series, I can easily do this with a computer. However, checking in the table of integrals, you will see that this equals ln 2. So you just wasted a few perfectly good hours to write and debug a program, instead of doing something fun. Not to mention that you are also wasting computer resources.

2. If the problem is more complicated, then the next question is: has anybody done this before? Almost always the answer will be "yes" and probably better. So it might be worth spending some time either searching the math libraries on your system or looking around on the internet. This does not mean you do not have to understand the problem or the algorithm used, on the contrary, you have to examine the code description carefully to see whether the particular solution will be adequate for you.

3. Okay, so you have finally convinced yourself that it must be done. Well, hold your horses, it is still too early to start typing your code in. Take a piece of paper and "design" your program. Write down the problem, the proposed solution and think about exceptions or areas where your solution will not be valid. For instance, make sure you will not get a division by zero or the log of a negative number. This will most likely involve checks at run time like $if(x! = 0.)y = a/x$, but it could be a lot more complicated. The best way to do this is with a flow chart and pseudo code. In making your flow chart it is not as important to use the standard symbols for ifs and loops, as it is to be consistent. Once you have made a flow chart, you can implement this in pseudo code, which merely phrases in English what you were trying to draw with the flow chart.

4. Once you have outlined your task, you can start your editor and write your program. Here the important thing to remember is that we usually forget very quickly what we have done a few weeks or, even worse, a month ago. If you ever want to use a program or function again later, or modify it, it is of the uttermost importance to comment as much as possible and use a naming convention, which lends itself to clarity. One-letter variables are generally **bad** because, first, you could only have a limited set and, second, the variables do not convey their meaning or usage.

## 3.9 Machine representation and precision

As a physicist trying to solve a problem or doing a measurement, you should always try to understand the quality or precision of your result. As you have learned in the introductory labs, if you quote a result you should include only as many digits as are significant where the least significant digit has been rounded off. So in an experiment it is your call to determine the precision based on your available instruments.

With computers you also have to be concerned with the precision of your calculation and with error propagation. Just because the computer printed out a result does not mean it is the correct one, even though a lot of people have ultimate confidence in computers and sometimes you hear statements like: *our computer has calculated this*. . . Of course apart from the rounding error we will discuss below, the program could also be just plain wrong.

One of the most important things to remember when you write computer code is:

> Every computer has a limit on how small or large a number can be.

Let us have a closer look at this statement. A computer internally represents numbers in binary form with so-called bits, where each bit represents a power of 2, depending on its position. You can think of these bits as switches, which are either open, representing a 0, or closed which means 1. If we take a hypothetical computer with 8 bits we could write a number in the following way

$$1 \quad 1 \quad 1 \quad 1 \quad 1 \quad 1 \quad 1 \quad 1$$
$$2^7 \quad 2^6 \quad 2^5 \quad 2^4 \quad 2^3 \quad 2^2 \quad 2^1 \quad 2^0$$

with the number 3 being represented by:

0000 0011

The highest number achievable on our computer would be $2^8 - 1$; i.e., all bits are set to 1. Clearly, if you now add one to this number the computer could not handle it and if you are lucky it will give you an error complaining about an overflow. An even more serious problem with our computer is that we do not have any negative numbers yet. The way this is handled by computers is to designate the bit all the way to the left, or the most significant bit, as the sign bit, leaving us with only 7 bits to express a number:

0111 1111

is now the highest number we can represent: $2^7 - 1 = 127$. As you might have guessed zero is written with all 0s:

0000 0000

The negative numbers are expressed in what is called the twos complement:

$+5 = 0000\ 0101$

$-5 = 1111\ 1011$

which you get by first forming the 1 complement (substituting a 1 for a 0 and vice versa) and then add 1. From this it is clear that the smallest number is

$1000\ 0000$

giving $2^7 = -128$.

In this way a computer only needs to have instructions for addition. For a Pentium III processor, which has 32 bit words the range is $[-2^{31}, 2^{31} - 1]$.

Now the world of science would be rather limited if we could only deal with integer numbers, so we also need floating point numbers. This is a little bit more complicated, but will also illustrate the problem with precision in more detail. The way to achieve a floating point number is to split up the 32 bits into three blocks, the most significant one still being the sign bit. The second block represents the exponent, and the third block is the mantissa. Here we show the number 0.5 represented as a so-called real number, the typical default 32 bit or 4 byte representation.

$$
\begin{array}{ccc}
0 & 1000\ 0000 & 100\ 0000\ 0000\ 0000\ 0000\ 0000 = 0.5 \\
\text{signbit} & \text{8-bit exponent} & \text{23-bit mantissa}
\end{array}
\tag{3.1}
$$

This can be expressed in the following way:

$$
x_{\text{float}} = (-1)^s * \text{mantissa} * 2^{\text{exp}-\text{bias}}
\tag{3.2}
$$

By expressing the exponent as an unsigned 8-bit word, we could only have positive exponents. In order to circumvent this rather drastic limitation, the exponent contains a so-called bias which is not written, but implicitly subtracted. This bias is 127, making the range of exponents $[0 - 127 = -127, 255 - 127 = 128]$. The most significant bit in the mantissa is the left-most bit, representing the value 1/2. From this discussion it is obvious that the precision of the machine is at the most:

$$
\frac{1}{2^{23}} = 1.2 * 10^{-7}
\tag{3.3}
$$

The reason this is the best case has to do with how the computer will add two floating point numbers. If you take the value 5 and you want to add $10^{-7}$ to it, you will still have only 5. In order to do this operation, the processor first has to make sure that the exponents for both numbers are the same, which means it has to shift the bits in the mantissa to the right until both exponents are the same. However, at $10^{-7}$ there are no places left for the bits to be right shifted, and the bits drop off; i.e., they are lost. You should always ask yourself what is the best achievable precision of your program and whether this is sufficient. To increase the precision you can work in double precision, using 64 bits or two words. This will give you an accuracy of $10^{-15}$, certainly good enough for most of your calculations.

## 3.10 Exercises

1. Make a subdirectory "Test" and create three files in your home directory, called "atest1," "btest2" and "ctest3." List the files alphabetically and in reverse order. Then move ctest3 to Test.
2. Still in your home directory, list the content of directory Test.
3. Create a .login file and define an alias so that if you type **lsd** it will list only directories (hint: use **grep**).
4. Insert a command for your .login file, which will tell you on which host you are logged in.
5. Delete all the files and directories you created in the first exercise.
6. Make a subdirectory in your home directory called "src."
7. Write a program which calculates the square root for any given number.
8. Go into the src directory and create a program which will calculate the area of a rectangle. Write the program in such a way that it asks you for input of the two sides and gives you the output. Compile and run it. Check for correctness.
9. Modify your program such that in case you input two equal sides, it will give you the radius of the outscribing circle.

# Chapter 4
# Interpolation

An important part in a scientist's life is the interpretation of measured data or theoretical calculations. Usually when you do a measurement you will have a discrete set of points representing your experiment. For simplicity, we assume your experiment to be represented by pairs of values: an independent variable "$x$," which you vary and a quantity "$y$," which is the measured value at the point $x$. As an illustration, consider a radioactive source and a detector, which counts the number of decays. In order to determine the half-life of this source, you would count the number of decays $N_0, N_1, N_2, \ldots, N_k$ at times $t_0, t_1, t_2, \ldots, t_k$. In this case $t$ would be your independent variable, which you hopefully would choose in such a way that it is suitable for your problem. However, what you measure is a discrete set of pairs of numbers $(t_k, N_k)$ in the range of $(t_0, t_k)$. In order to extract information from such an experiment, we would like to be able to find an analytical function which would give us $N$ for any arbitrary chosen point $t$. But, sometimes trying to find an analytical function is impossible, or even though the function might be known, it is too time consuming to calculate or we might be only interested in a small local region of the independent variable.

To illustrate this point, assume your radioactive source is $^{241}$Am, an $\alpha$ emmiter. Its half-life is $\tau_{1/2} = 430$ years. Clearly you cannot determine the half-life by measuring it. Because it is very slowly decaying you probably will measure the activity over a longer time period, say every Monday for a couple of months. After five months you would stop and look at the data. One question you might want to answer is: what was the activity on Wednesday of the third week? Because this day is inside your range of $(t_0, t_k)$ you would use interpolation techniques to determine this value. If, on the other hand, you want to know the activity eight months from the end of your measurement, you would extrapolate to this point from the previous series of measurements. The idea of interpolation is to select a function $g(x)$ such that $g(x_i) = f_i$ for each data point $i$ and that this function is a good approximation for any other $x$ lying between the original data points. But what can we consider as a good approximation to the original data if we do not have the original function? Because data points may be interpolated by an infinite number of functions we should have some criterion or a guideline to select a reasonable function. In mathematics there are very many theorems on function analysis including interpolation with error analysis. As a rule these methods are grounded on "smoothness" of the interpolated functions. But this would not work for functions such as the example given by Runge of $1/(1 + 25x^2)$ on the interval $[-1, +1]$. Before we go into the discussion of interpolation techniques, we need to add a word of caution. Because you measure discrete points, you have to be very careful about the spacing of your independent variable. If these points are too far apart, you will loose information in between and your prediction from interpolation would be totally off. Figure 4.1 illustrates this point. Assuming you have made six measurements at the indicated points, you will clearly miss the oscillatory behavior of the function. In addition, judging from the points and taking into account the error bars, a straight line is probably what you would assume for the function's behavior.

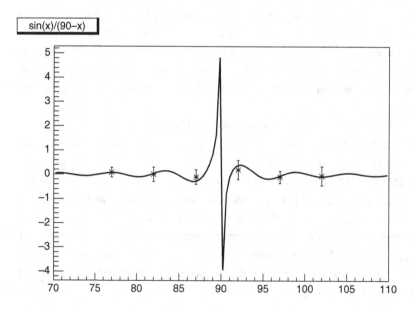

**Figure 4.1** An example to illustrate the dangers of interpolation.

## 4.1 Lagrange interpolation

As a first step into interpolation we look at Lagrange interpolation. This will help us to understand the principles of more complicated interpolation methods like Neville's algorithm. The method relies on the fact that in a finite interval $[a,b]$ a function $f(x)$ can always be represented by a polynomial $P(x)$. Our task will be to find this polynomial $P(x)$, from the set of points $(x_i, f(x_i))$. If we have just two such pairs, the interpolation is straightforward and we are sure you have used it in some lab experiment before, looking up tabulated values and determining a point in between two listed values.

Let us take a look at the vapor pressure of $^4$He as a function of temperature (Figure 4.2). In the literature you would find tabulated values like this

| Temperature [K] | Vapor pressure [kPa] |
| --- | --- |
| 2.3 | 6.38512 |
| 2.7 | 13.6218 |
| 2.9 | 18.676 |
| 3.2 | 28.2599 |
| 3.5 | 40.4082 |
| 3.7 | 49.9945 |

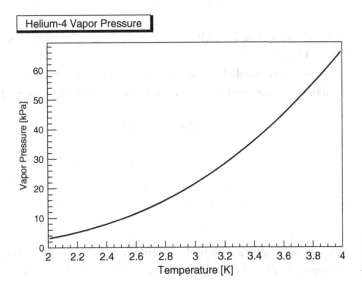

**Figure 4.2** $^4$He vapor pressure as a function of temperature.

and your task is to find the pressure at 3.0 K. The most straightforward way is to do a linear interpolation between the two points. You would set up two equations

$$b = \frac{y_2(x_2) - y_1(x_1)}{x_2 - x_1} \tag{4.1}$$

$$a = y_1(x_1) - x_1 \frac{y_2(x_2) - y_1(x_1)}{x_2 - x_1} \tag{4.2}$$

in order to solve:

$$y(x) = a + b * x \tag{4.3}$$

Combining equations (4.1) and (4.2) and some reshuffling gives

$$y(x) = \frac{x - x_2}{x_1 - x_2} y_1(x_1) + \frac{x - x_1}{x_2 - x_1} y_2(x_2) \tag{4.4}$$

which is the equation for a line through $(x_1, y_1(x_1))$ and $(x_2, y_2(x_2))$. Using this interpolation we get a value of 21.871 kPa at 3 K, while the parameterization gives 21.595 kPa.

In order to improve our result we could use a second degree polynomial and employ a quadratic interpolation. In this case our interpolation function would be:

$$y(x) = \frac{(x - x_2)(x - x_3)}{(x_1 - x_2)(x_1 - x_3)} y_1(x_1) + \frac{(x - x_1)(x - x_3)}{(x_2 - x_1)(x_2 - x_3)} y_2(x_2)$$
$$+ \frac{(x - x_1)(x - x_2)}{(x_3 - x_1)(x_3 - x_2)} y_3(x_3) \tag{4.5}$$

Using the points at 2.7, 2.9 and 3.2 K, our vapor pressure for 3 K is now 21.671 kPa.

The next step would be to use four points and construct a third degree polynomial. In general, we can write for the interpolation in terms of a polynomial

$$P(x) = \sum_{k=1}^{N} \lambda_k(x) f(x_k) \tag{4.6}$$

where

$$\lambda_k(x) = \frac{\prod_{l=1 \neq k}^{N} (x - x_l)}{\prod_{l=1 \neq k}^{N} (x_k - x_l)} \tag{4.7}$$

which is the Lagrange interpolation formula. This is a polynomial which has the values $y_k$ $(k = 1, 2, 3, \ldots, N)$ for $x_k$ $(k = 1, 2, 3, \ldots, N)$.

## 4.2 Neville's algorithm

Neville's algorithm provides a good way to find the interpolating polynomial. In this method we are using linear interpolations between successive iterations. The degree of the polynomial will be just the number of iterations we have done. Suppose you have five measurements of the vapor pressure $P_i(T_i)$ at five different temperatures $T_i$ and you would like to find the vapor pressure for an intermediate temperature $T$. The first iteration is to determine a linear polynomial $P_{ij}$ between neighboring points for all the values we have:

$$P_{12} = \frac{1}{T_1 - T_2}((T - T_2)P(T_1) - (T - T_1)P(T_2)) \tag{4.8}$$

$$P_{23} = \frac{1}{T_2 - T_3}((T - T_3)P(T_2) - (T - T_2)P(T_3)) \tag{4.9}$$

$$P_{34} = \frac{1}{T_3 - T_4}((T - T_4)P(T_3) - (T - T_3)P(T_4)) \tag{4.10}$$

$$P_{45} = \frac{1}{T_4 - T_5}((T - T_5)P(T_4) - (T - T_4)P(T_5)) \tag{4.11}$$

The next iteration will again be an interpolation but now between these intermediate points:

$$P_{123} = \frac{1}{T_1 - T_3}((T - T_3)P_{12} - (T - T_1)P_{23}) \tag{4.12}$$

$$P_{234} = \frac{1}{T_2 - T_4}((T - T_4)P_{23} - (T - T_2)P_{34}) \tag{4.13}$$

$$P_{345} = \frac{1}{T_3 - T_5}((T - T_5)P_{34} - (T - T_3)P_{45}) \tag{4.14}$$

Our interpolation is now a polynomial of degree two. We will continue this process for two more steps to end up with a fourth degree polynomial $P_{12345}$ which is our final point. By including more points in your interpolation you would increase the degree of your polynomial. The following diagram helps to visualize this procedure:

$$
\begin{array}{ccccc}
T_1 & P(T_1) & & & \\
 & & P_{12} & & \\
T_2 & P(T_2) & & P_{123} & \\
 & & P_{23} & & P_{1234} \\
T_3 & P(T_3) & & P_{234} & & P_{12345} \\
 & & P_{34} & & P_{2345} \\
T_4 & P(T_4) & & P_{345} & \\
 & & P_{45} & & \\
T_5 & P(T_5) & & & \\
\end{array}
$$

## 4.3 Linear interpolation

The idea of linear interpolation is to approximate data at a point $x$ by a straight line passing through two data points $x_j$ and $x_{j+1}$ closest to $x$:

$$g(x) = a_0 + a_1 x \tag{4.15}$$

where $a_0$ and $a_1$ are coefficients of the linear functions. The coefficients can be found from a system of equations

$$g(x_j) = f_j = a_0 + a_1 x_j \tag{4.16}$$

$$g(x_{j+1}) = f_{j+1} = a_0 + a_1 x_{j+1} \tag{4.17}$$

Solving this system for $a_0$ and $a_1$ the function $g(x)$ takes the form

$$g(x) = f_j + \frac{x - x_j}{x_{j+1} - x_j}\left(f_{j+1} - f_j\right) \tag{4.18}$$

on the interval $[x_j, x_{j+1}]$.

It is clear that for the best accuracy we have to pick points $x_j$ and $x_{j+1}$ closest to $x$. The linear interpolation (4.18) also may be written in a symmetrical form

$$g(x) = f_j \frac{x - x_{j+1}}{x_j - x_{j+1}} + f_{j+1} \frac{x - x_j}{x_{j+1} - x_j} \tag{4.19}$$

Application of the linear interpolation for $f(x) = \sin(x^2)$ is shown in Figure 4.3. We selected ten equidistant data points to present the original

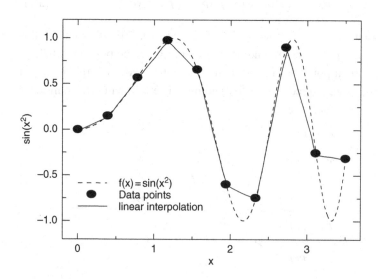

**Figure 4.3** Linear interpolation for $f(x) = \sin(x^2)$.

function on the interval [0.0, 3.5]. Then the linear interpolation is applied to the each $x_j$ and $x_{j+1}$ interval. From the figure we may draw a conclusion that the linear interpolation may work well for very smooth functions when the second and higher derivatives are small. We may improve the quality of linear interpolation by increasing the number of data points $x_i$ on the interval.

It is worth noting that for each data interval one has a different set of coefficients, $a_0$ and $a_1$. This is the principal difference from data fitting where the same function, with the same coefficients, is used to fit the data points on the whole interval $[x_1, x_n]$.

## 4.4 Polynomial interpolation

Polynomial interpolation is a very popular method due, in part, to its simplicity

$$g(x) = a_0 + a_1 x + a_2 x^2 + \cdots + a_n x^n \qquad (4.20)$$

The condition that the polynomial $g(x)$ passes through sample points $f_j(x_j)$

$$f_j(x_j) = g(x_j) = a_0 + a_1 x_j + a_2 x_j^2 + \cdots + a_n x_j^n \qquad (4.21)$$

generates a system of $n+1$ linear equations to determine coefficients $a_j$. The number of data points minus one that are used for interpolation defines the *order of interpolation*. Thus, linear (or two-point) interpolation is the first order interpolation.

In Chapter 9 we discuss how to solve a system of linear equations. However, there is another way of obtaining coefficients for polynomial interpolation. Let us consider three-point (or second order) interpolation for three given points $x_j$, $f_j$ with $j, j+1, j+2$:

$$f_j = a_0 + a_1 x_j + a_2 x_j^2$$
$$f_{j+1} = a_0 + a_1 x_{j+1} + a_2 x_{j+1}^2 \qquad (4.22)$$
$$f_{j+2} = a_0 + a_1 x_{j+2} + a_2 x_{j+2}^2$$

This system may be solved analytically for the coefficients $a_0, a_1, a_2$. Substituting coefficients $a_j$ into equation (4.20), the interpolated function $g(x)$ may be written in the following symmetrical form

$$g(x) = f_j \frac{(x - x_{j+1})(x - x_{j+2})}{(x_j - x_{j+1})(x_j - x_{j+2})} + f_{j+1} \frac{(x - x_j)(x - x_{j+2})}{(x_{j+1} - x_j)(x_{j+1} - x_{j+2})}$$
$$+ f_{j+2} \frac{(x - x_j)(x - x_{j+1})}{(x_{j+2} - x_j)(x_{j+2} - x_{j+1})} \qquad (4.23)$$

Comparing this result with equation (4.19) for linear interpolation one may write easily the interpolating polynomial of degree $n$ through $n+1$ points

$$g(x) = f_1 \frac{(x-x_2)(x-x_3)\cdots(x-x_{n+1})}{(x_1-x_2)(x_1-x_3)\cdots(x_1-x_{n+1})}$$
$$+ f_2 \frac{(x-x_1)(x-x_3)\cdots(x-x_{n+1})}{(x_2-x_1)(x_2-x_3)\cdots(x_2-x_{n+1})}$$
$$+ \cdots + f_{n+1} \frac{(x-x_1)(x-x_2)\cdots(x-x_n)}{(x_{n+1}-x_1)(x_{n+1}-x_2)\cdots(x_{n+1}-x_n)} \tag{4.24}$$

This is Lagrange's classical formula for polynomial interpolation.

In Figure 4.4 you can see application of first, third, fifth and seventh order polynomial interpolation to the same data set. It is clear that moving from the first order to the third and fifth order improves interpolated values to the original function. However, the seventh order interpolation instead of being closer to the function $f(x)$ produces wild oscillations. This situation is not uncommon for high order polynomial interpolation. Thus the apparent simplicity of the polynomial interpolation has a down side. There is a rule of thumb: do not use high order interpolation. Fifth order may be considered as a practical limit. If you believe that the accuracy of the fifth order interpolation is not sufficient, then you should consider some other method of interpolation.

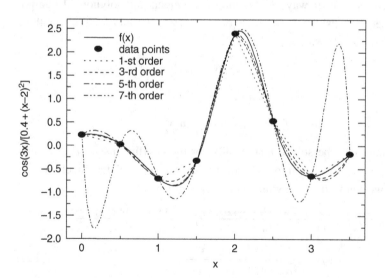

**Figure 4.4** Polynomial interpolation.

## 4.5 Cubic spline

One of the principal drawbacks of the polynomial interpolation is related to the discontinuity of derivatives at data points $x_j$. One way to improve polynomial interpolation considerably is the spline interpolation.

The procedure for deriving coefficients of spline interpolations uses information from all data points, i.e., nonlocal information, to guarantee global smoothness in the interpolated function up to some order of derivatives.

The idea of spline interpolation is reminiscent of very old mechanical devices used by draftsmen to obtain a smooth shape. It is like securing a strip of elastic material (metal or plastic ruler) between knots (or nails). The final shape is quite smooth (Figure 4.5).

Cubic splines are the most popular method. In this case the interpolated function on the interval $(x_j, x_{j+1})$ is presented in the form

$$g(x) = f_j + b_j(x - x_j) + c_j(x - x_j)^2 + d_j(x - x_j)^3 \qquad (4.25)$$

For each interval we need to have a set of three parameters: $b_j$, $c_j$ and $d_j$. Because there are $(n - 1)$ intervals, one has to have $3n - 3$ equations for deriving the coefficients for $j = 1, \ldots, n - 1$. The fact that $g_j(x_j) = f_j(x_j)$ imposes $(n - 1)$ equations. Central to spline interpolation is the idea that the

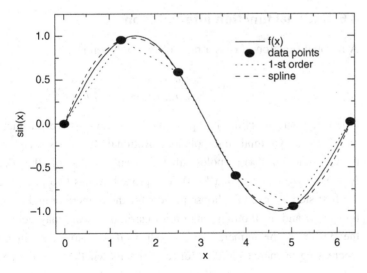

**Figure 4.5** Spline interpolation.

interpolated function $g(x)$ has continuous first and second derivatives at each of the $n-2$ interior points $x_j$, i.e.:

$$g'_{j-1}x_j = g'_j x_j \qquad (4.26)$$

$$g''_{j-1}x_j = g''_j x_j \qquad (4.27)$$

These conditions impose $2(n-2)$ equations resulting in $n-1+2(n-2) = 3n-5$ equations for the coefficients. Thus two more conditions are needed. There are a few possibilities to fix two more conditions, for example, for natural splines the second order derivatives are zero on boundaries. Solving analytically a system with many equations is straightforward but cumbersome. However, computers do this job very fast and efficiently. Because there are many programs for cubic interpolation available, we recommend you use one of those, instead of writing your own spline program. In Appendix B.1 and B.2 you will find a list of math libraries with spline programs, available on the Web.

Generally, spline does not have advantages over polynomial interpolation when used for smooth, well behaved data, or when data points are close on the $x$ scale. The advantage of spline comes into the picture when dealing with "sparse" data, when there are only a few points for smooth functions or when the number of points is close to the number of expected maximums. From Figure 4.5 you can see how well spline interpolation fits $f(x) = \sin(x)$ on only six(!) data points.

## 4.6 Rational function interpolation

A rational function $g(x)$ is a ratio of two polynomials:

$$g(x) = \frac{a_0 + a_1 x + a_2 x^2 + \cdots + a_n x^n}{b_0 + b_1 x + b_2 x^2 + \cdots + b_m x^m} \qquad (4.28)$$

Often rational function interpolation is a more powerful method compared to polynomial interpolation. Rational functions may well interpolate functions with poles, that is with zeros of the denominator $b_0 + b_1 x + b_2 x^2 + \cdots + b_m x^m = 0$. The procedure has two principal steps. In the first step we need to choose powers for the numerator and the denominator, i.e., $n$ and $m$. Plotting data on a logarithmic scale may help to evaluate the power of the numerator for small $x$, if we have data in this region. Decreasing or increasing data for large $x$ tells whether $n < m$ or vice versa.

It is difficult to say what would be the best combination of the powers $n$ and $m$. You may need a few trials before coming to a conclusion. For example, our analysis tells us that the degree of the numerator is $n = 2$ and the degree of the denominator is $m = 1$. Then:

$$g(x) = \frac{a_0 + a_1 x + a_2 x^2}{b_0 + b_1 x} \tag{4.29}$$

Now we know the number of parameters we need to find is five. However, we should fix one of the coefficients because only the ratio makes sense. If we choose, for example, $b_0$ as a fixed number $c$, then we need four data points to solve a system of equations to find the coefficients

$$f(x_1)(c + b_1 x_1) = a_0 + a_1 x_1 + a_2 x_1^2$$

$$f(x_2)(c + b_1 x_2) = a_0 + a_1 x_2 + a_2 x_2^2$$

$$f(x_3)(c + b_1 x_3) = a_0 + a_1 x_3 + a_2 x_3^2 \tag{4.30}$$

$$f(x_4)(c + b_1 x_4) = a_0 + a_1 x_4 + a_2 x_4^2$$

After a few rearrangements the system may be written in the traditional form for a system of linear equations. Very stable and robust programs for rational function interpolation can be found in many standard numerical libraries.

## 4.7 Exercises

1. Write a program that implements the first order (linear) interpolations.
2. Write a program that implements $n$-point Lagrange interpolation. Treat $n$ as an input parameter.
3. Apply the program to study the quality of the interpolation to functions $f(x) = \sin(x^2)$, $f(x) = e^{\sin(x)}$ and $f(x) = 0.2/[(x - 3.2)^2 + 0.04]$ initially calculated in 10 uniform points in the interval $[0.0, 5.0]$. Compare the results with the cubic spline interpolation.
4. Use third and seventh order polynomial interpolation to interpolate Runge's function:

$$f(x) = \frac{1}{1 + 25x^2}$$

at the $n = 11$ points $x_i = -1.0, -0.8, \ldots, 0.8, 1.0$. Compare the results with the cubic spline interpolation.
5. Study how the number of data points for interpolation affects the quality of interpolation of Runge's function in the example above.

# Chapter 5
# Taking derivatives

In physics, one of the most basic mathematical tasks is differentiation. The first laws of physics you encountered were Newton's laws, where the first and second laws are differential equations relating derivatives to a function. Another every day life derivative is velocity, and there are many more you encounter on a daily basis. In this chapter we will discuss taking derivatives numerically.

## 5.1 General discussion of derivatives with computers

The most important statement one can make about taking derivatives with computers is:

$$\boxed{\textbf{Don't}}$$

if you can avoid it.

To illuminate this rather drastic statement, let us start with a function $f(x)$ and the definition of the derivatives with respect to $x$:

$$\frac{df(x)}{dx} = \lim_{h \to 0} \frac{f(x+h) - f(x)}{h} \tag{5.1}$$

There are two issues with this equation: the fact that this is defined for $h \to 0$ and that you subtract two numbers from each other which only differ by a small amount. Clearly the second point is easy to address. As we have seen in Chapter 3, the computer has only a finite representation and when your difference is smaller than $10^{-7}$ this is then for the computer equal to zero. However, be careful; the standard for C and C++ is to convert float into double for any arithmetic operation. As is pointed out in *Numerical Recipes* [5] this implicit conversion is "sheer madness." The result still will come out as a float, but now using much longer CPU time because the natural processor word is 32 bit and implicitly you calculate in 64 bit.

## 5.2 Forward difference

Let us now start with the first example of taking the derivative numerically, the so-called forward derivative. The starting point for deriving this method is, as in many other cases as well, the Taylor series. Any function $f(x)$ can be written as a Taylor series around $x + h$:

$$f(x+h) = f(x) + \frac{h}{1!}f'(x) + \frac{h^2}{2!}f''(x) + \frac{h^3}{3!}f'''(x) + \cdots \qquad (5.2)$$

We can solve this equation for $f'(x)$ which results in:

$$f'(x) = \frac{1}{h}\left(f(x+h) - f(x) - \frac{h^2}{2!}f''(x) - \frac{h^3}{3!}f'''(x) - \cdots\right) \qquad (5.3)$$

or in a more intuitive form

$$f'(x) = \frac{f(x+h) - f(x)}{h} - \left(\frac{h}{2!}f''(x) + \frac{h^2}{3!}f'''(x) + \cdots\right) \qquad (5.4)$$

As you can see, the first term is very close to our definition of the derivative, but the $h \to 0$ is missing. This is the term which we are going to use as our approximation, and in doing so we are introducing an error of order $h$. Question: in what case do we have a smaller error? So our forward difference derivative can be written as:

$$f'(x) \approx \frac{f(x+h) - f(x)}{h} \qquad (5.5)$$

You can display this graphically as approximating the function with a straight line between two $x$-values. Clearly we have chosen too large a step, but this helps in illustrating the point (Figure 5.1).

## 5.3 Central difference and higher order methods

A much improved version can be derived by using an additional point leading to a three-point formula. We again start with the Taylor series, but instead of using only $x$ and $x + h$, we will also employ $x - h$. This means we will expand our function around $f(x)$ for $f(x+h)$ and $f(x-h)$. This leads to two equations:

$$f(x+h) = f(x) + \frac{h}{1!}f'(x) + \frac{h^2}{2!}f''(x) + \frac{h^3}{3!}f'''(x) + \cdots \qquad (5.6)$$

$$f(x-h) = f(x) - \frac{h}{1!}f'(x) + \frac{h^2}{2!}f''(x) - \frac{h^3}{3!}f'''(x) + \cdots \qquad (5.7)$$

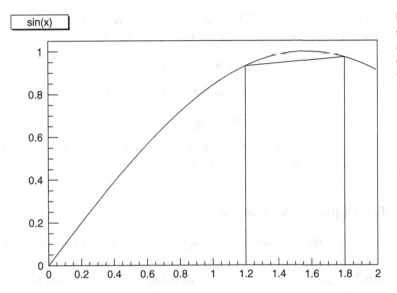

We still want to solve for $f'(x)$ but, first we subtract equation (5.6) from (5.7) to get:

$$f(x+h) - f(x-h) = 2\frac{h}{1!}f'(x) + 2\frac{h^3}{3!}f'''(x) + \text{odd terms} \qquad (5.8)$$

which then leads to

$$f'(x) = \frac{f(x+h) - f(x-h)}{2h} + \mathcal{O}(h^2) \qquad (5.9)$$

where $\mathcal{O}(h^2)$ stands for the lowest order of the remainder. The advantage of this method is that all the terms odd in $h$ drop out and we have increased the accuracy by one order.

Clearly we do not have to stop here. Instead of one point on each side of $x$, we can take several, for instance at $x - 2h, x - h, x + h$, and $x + 2h$. Instead of a three-point formula we are now using a five-point formula which improves the accuracy now to $\mathcal{O}(h^4)$ shown in equation (5.10):

$$f'(x) = \frac{1}{12h}(f(x-2h) - 8f(x-h) + 8f(x+h) - f(x+2h)) + \mathcal{O}(h^4) \qquad (5.10)$$

However, from the point of view of computational methods, two remarks are necessary. First, if the function is not known outside some boundaries, but you have to determine the derivative at the boundaries, you have to extrapolate either the function or the derivative. Second, we have written equation (5.10) grouped together in a mathematically intuitive way. For computational purposes it is better to do the following:

$$f_1 = f(x - 2h) + 8f(x + h)$$

$$f_2 = 8f(x - h) + f(x + 2h)$$

$$f'(x) = \frac{1}{12h}(f_1 - f_2)$$

In this way you reduce the number of subtractions by one, therefore reducing the problem of subtractive cancellation.

## 5.4 Higher order derivatives

A closer look at equations (5.6) and (5.7) also reveals the way to get the second derivative. Instead of subtracting them we will add the two equations, therefore getting rid of $f'(x)$ and calculating $f''(x)$. This leads to:

$$f''(x) = \frac{f(x + h) + f(x - h) - 2f(x)}{h^2} + \mathcal{O}(h^2) \qquad (5.11)$$

which is a three-point formula for the second derivative.

## 5.5 Exercises

1. Write a program to calculate sin and cos and determine the forward differentiation.
2. Do the same, but use central difference.
3. Plot all derivatives and compare with the analytical derivative.
4. The half-life $t_{1/2}$ of $^{60}$Co is 5.271 years. Write a program which calculates the activity as a function of time and amount of material. Design your program in such a way that you could also input different radioactive materials.
5. Write a program which will calculate the first and second derivatives for any function you give.

# Chapter 6
# Numerical integration

## 6.1 Introduction to numerical integration

In the previous chapter we discussed derivatives and their importance in physics. Now we need to address the inverse, namely integration. A lot of problems in physics are described by derivatives, for example the free fall. In a real-world experiment you would be standing on top of a large tower with a stop watch and you would drop an object, determining the acceleration $g$ from the time it takes the object to hit the ground and the known height $h$ of the tower. This clearly involves two integrations:

$$v_f = \int_0^t g \, dt \tag{6.1}$$

$$h = \int_0^t gt \, dt \tag{6.2}$$

This of course is a problem you would not consider solving numerically, but it illustrates the need for integration in physics. More complicated problems involve determining the normalization constant for a wavefunction, which you find from:

$$\int_{-\infty}^{+\infty} |\Psi(x)|^2 \, dx = 1 \tag{6.3}$$

Again, it is helpful to look at the definition of an integral, especially the case of definite integrals, which we will be dealing with. According to Riemann[1] a function is integrable if the limit $I$

$$\int_a^b f(x) \, dx = \lim_{\Delta x \to 0} \sum_{i=0}^{n-1} f(x_i) \Delta x_i \tag{6.4}$$

---

[1] Bernhard Riemann, 1826–1866, German mathematician

exists, where the interval has been divided into $n-1$ slices. Again, we see the brute force approach clearly: let us just take the computer, divide this interval into many slices with a small $\Delta x$ and be done with it. This is pretty much what we will do in the next chapter, and for a quick and "dirty" answer it is just fine.

## 6.2 The simplest integration methods

### The rectangular and trapezoid integration

First we start with the rectangular method. The idea behind this is to divide the interval $[a, b]$ into $N$ intervals, equally spaced by $h = x_i - x_{i-1}$, and approximate the area under the curve by simple rectangular strips. The width of the strips will be $h$, while the height is given by the average of the function in this interval. We could take:

$$\overline{f_i(x)} = \frac{f_i(x) + f_{i-1}(x)}{2} \tag{6.5}$$

Or to speed up the calculation, we could use the fact that for a slowly varying function the value $f(x_{i-1/2})$ at $x_i - h/2$ is a good approximation to $\overline{f_i(x)}$. In this case our integral is expressed as:

$$\int_a^b f(x)\, dx = h \sum_{t=1}^N f_{i-1/2} \tag{6.6}$$

which is known as the rectangular method (Figure 6.1). Clearly, the smaller you make the width of the rectangles the smaller the error becomes, as long as you are not dominated by the finite machine representation. Naturally the price you pay is increased computational time.

A better approach is to employ a trapezoid for the areas which will represent the integral of the function. We are still using $N$ slices, but instead of using the function value in the middle, we are now determining the area by using a trapezoid, which consists of the four points $(x_i, 0)$, $(x_i, f(x_i))$, $(x_{i+1}, f(x_{i+1}))$, $(x_{i+1}, 0)$. The area of this trapezoid is then:

$$I_{[x_i, x_{i+1}]} = h\frac{f(x_i) + f(x_{i+1})}{2} \tag{6.7}$$

$$= \frac{h}{2}(f(x_i) + f(x_{i+1}))$$

One of the things to note here is that the values inside the boundaries are used twice, while the function values at the boundaries $a$ and $b$ are each used only once. Keeping this in mind, now we can write the trapezoidal rule as:

$$\int_a^b f(x)\, dx = h\left(\frac{1}{2}f(x_0 = a) + f(x_0 + h) + \cdots + \frac{1}{2}f(x_N = b)\right) \tag{6.8}$$

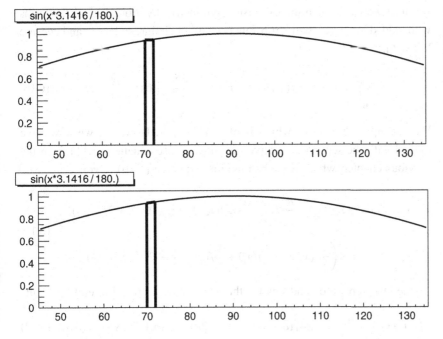

**Figure 6.1** The integral of sin(x) with the rectangular (top) and trapezoidal (bottom) methods.

## The Simpson method

In Simpson's[2] method we are approximating the function in each interval by a parabola. To derive the method we are going back to Taylor and doing a Taylor expansion of the integral around the midpoint $x_i$ in an interval $[x_{i-1}, x_{i+1}]$. This looks rather complicated at first, but the reason for this will become immediately clear once we have a closer look at the expansion:

$$\int_{x_{i-1}}^{x_{i+1}} f(x)\, dx = f(x_i) \int_{x_{i-1}}^{x_{i+1}} dx + \frac{f'(x_i)}{1!} \int_{x_{i-1}}^{x_{i+1}} (x - x_i)\, dx$$

$$+ \frac{f''(x_i)}{2!} \int_{x_{i-1}}^{x_{i+1}} (x - x_i)^2\, dx + \frac{f'''(x_i)}{3!} \int_{x_{i-1}}^{x_{i+1}} (x - x_i)^3\, dx \qquad (6.9)$$

$$+ \cdots + \frac{f^{(n)}(x_i)}{n!} \int_{x_{i-1}}^{x_{i+1}} (x - x_i)^n\, dx$$

---

[2] Thomas Simpson, 1710–1761, English mathematician.

Because we have taken the expansion symmetrically around $x_i$ all the terms with odd derivatives will drop out because their integral is 0 and we are left with:

$$\int_{x_{i-1}}^{x_{i+1}} f(x)\,dx = f(x_i)*2h + f''(x_i)*\frac{2h^3}{3*2!} + \mathcal{O}(h^5 f^{(4)}(x_i)) \qquad (6.10)$$

We end up with a result which is of order $h^5 f^{(4)}$. However, we now have to deal with the second derivative. For this we use equation (5.11) from the previous chapter, which we substitute into equation (6.10) to get:

$$\int_{x_{i-1}}^{x_{i+1}} f(x)\,dx = 2hf(x_i) + \frac{h}{3}(f(x_{i-1}) - 2f(x_i) + f(x_{i+1})) + \mathcal{O}(h^5 f^{(4)}(x_i))$$

$$= h\left(\frac{1}{3}f(x_{i-1}) + \frac{4}{3}f(x_i) + \frac{1}{3}f(x_{i+1})\right) + \mathcal{O}(h^5 f^{(4)})(x_i)) \qquad (6.11)$$

As the last step we extend this for the whole interval $[a, b]$ to end up with:

$$\int_a^b f(x)\,dx = \frac{h}{3}(f(a) + 4f(a+h) + 2f(a+2h) + 4f(a+3h) + \cdots + f(b)) \qquad (6.12)$$

Remember, we are using *three* points for our integration. This requires that the number of intervals is even and therefore the number of points is odd.

## 6.3 More advanced integration

### Gaussian integration with Legendre polynomials

We will now turn our attention to a more sophisticated technique of integration. All the previous methods can be expressed with the following general approximation:

$$\int_a^b f(x)\,dx = \sum_{i=1}^{N} f(x_i)w_i \qquad (6.13)$$

where $w_i$ is the corresponding weight for $f(x_i)$. In the case of the trapezoidal method the weights are $(h/2, h/2)$, and in the case of the Simpson method the weights are $(h/3, 4h/3, h/3)$. In addition, the points $x_i$ are separated equidistantly. These integrations are also known as quadrature, and we can consider them as approximations of the integrands by polynomials in each slice. The next form, Gauss–Legendre integration, will approximate the integrand with a polynomial over the whole interval. The fundamental idea behind this method lies in the fact that we can express a function $f(x)$ in an interval $[a, b]$ in terms of a complete set of polynomials $P_l$,

$$f(x) = \sum_{l=0}^{n} \alpha_l P_l \qquad (6.14)$$

As you will see below, it is convenient to choose orthogonal polynomials, and the first such set we will use are the Legendre[3] polynomials, which have the following property:

$$\int_{-1}^{+1} P_n(x) P_m(x) = \frac{2}{2m+1} \delta_{n,m} \qquad (6.15)$$

Because these polynomials are only defined in $[-1, 1]$, we have to change our integration limits from $[a, b] \rightarrow [-1, 1]$ by the following substitution:

$$x \rightarrow \frac{b-a}{2} x + \frac{b+a}{2} \qquad (6.16)$$

leading to

$$\int_{a}^{b} f(x)\, \mathrm{d}x \rightarrow \frac{b-a}{2} \int_{-1}^{+1} f\left(\frac{b-a}{2} x + \frac{b+a}{2}\right) \mathrm{d}x \qquad (6.17)$$

For some functions, other polynomials will be more convenient, and you will have to remap your integral limits to those for which the polynomials are defined. Because this remapping is trivial, for the following discussion, we will assume that we are only interested in the integral in $[-1, 1]$.

As the starting point for our discussion we will assume that the function $f(x)$ can be approximated by a polynomial of order $2n - 1$, $f(x) \approx p_{2n-1}(x)$. If this is a good representation we can express our integral with:

$$\int_{-1}^{+1} f(x)\, \mathrm{d}x = \sum_{i=1}^{n} w_i f(x_i) \qquad (6.18)$$

In other words, knowing the $n$ abscissas $x_i$ and the corresponding weight values $w_i$ we can compute the integral. This means that we have to find out how to determine these quantities. In the first step we will decompose $p_{2n-1}$ into two terms

$$p_{2n-1}(x) = p_{n-1}(x) P_n(x) + r_{n-1} \qquad (6.19)$$

---

[3] Adrien-Marie Legendre, 1752–1833, French mathematician.

where $P_n$ is a Legendre polynomial of order $n$, while $p_{n-1}$ and $r_{n-1}$ are polynomials of order $(n-1)$. Expressing $p_{n-1}$ by Legendre polynomials as well, we get

$$\int_{-1}^{+1} p_{2n-1}(x)\,\mathrm{d}x = \sum_{k=0}^{n-1} \alpha_k \int_{-1}^{+1} P_k(x)P_n(x)\,\mathrm{d}x + \int_{-1}^{+1} r_{n-1}(x)\,\mathrm{d}x \qquad (6.20)$$

Because of the orthogonality, the first term vanishes and we are left with:

$$\int_{-1}^{+1} p_{2n-1}(x)\,\mathrm{d}x = \int_{-1}^{+1} r_{n-1}(x)\,\mathrm{d}x \qquad (6.21)$$

Furthermore, we know that $P_n$ has $n$ zeros in $[-1, +1]$, which we will label $x_1, x_2, \ldots, x_n$. For these points we have the relation:

$$p_{2n-1}(x_i) = r_{2n-1}(x_i) \qquad (6.22)$$

Expressing $r_{n-1}(x)$ with Legendre polynomials

$$r_{n-1}(x) = \sum_{k=0}^{n-1} w_k P_k(x) \qquad (6.23)$$

and using equation (6.22) we finally get:

$$p_{2n-1}(x_i) = \sum_{k=0}^{n-1} w_k P_k(x_i) \qquad (6.24)$$

The $P_k(x_i)$ are the values of the Legendre polynomials of order $k$, up to $k = n - 1$ evaluated at the roots $x_i$ of the Legendre polynomial $P_n(x)$. If we write this out for $k = 2$ we get:

$$p_{2n-1}(x_1) = w_0 P_0(x_1) + w_1 P_1(x_1) + w_2 P_2(x_1)$$
$$p_{2n-1}(x_2) = w_0 P_0(x_2) + w_1 P_1(x_2) + w_2 P_2(x_2)$$
$$p_{2n-1}(x_3) = w_0 P_0(x_3) + w_1 P_1(x_3) + w_2 P_2(x_3)$$

which we can write conveniently in matrix form:

$$p_{2n_1}(x_1, \ldots, x_n) = \begin{pmatrix} P_0(x_1) & P_1(x_1) & \ldots & P_k(x_1) \\ \cdots\cdots\cdots\cdots\cdots\cdots\cdots \\ P_0(x_n) & P_1(x_n) & \cdots & P_k(x_n) \end{pmatrix} \begin{pmatrix} w_0 \\ w_1 \\ \\ w_k \end{pmatrix} \qquad (6.25)$$

In order to solve for the unknown $w_k$ we have to invert the matrix $\mathbf{P}$:

$$w_k = \sum_{i=1}^{n} p_{2n-1}(x_i)\{\mathbf{P}^{-1}\}_{ik} \qquad (6.26)$$

Finally, after all this mathematical hoopla we get back to our integral:

$$\int_{-1}^{+1} f(x) \, dx - \int_{-1}^{+1} p_{2n-1}(x) \, dx = \sum_{k=0}^{n-1} w_k \int_{-1}^{+1} P_k(x) \, dx \qquad (6.27)$$

Using the fact that $P_{k=0}(x) = 1$, we can multiply this equation with $P_0(x)$ and use the orthogonality condition to get

$$\int_{-1}^{+1} p_{2n-1}(x) \, dx = 2w_0 = 2\sum_{i=1}^{n} p_{2n-1}(x_i)\{\mathbf{P}^{-1}\}_{i0} \qquad (6.28)$$

since

$$\int_{-1}^{+1} P_k(x)P_0(x) \, dx = \frac{2}{2k+1}\delta_{k,0} \qquad (6.29)$$

from the properties of the Legendre polynomials. Up to now we have assumed that $f(x)$ is exactly represented by a polynomial, while in most cases this is an approximation:

$$p_{2n-1}(x_i) \approx f(x_i) \qquad (6.30)$$

Using equation (6.30) we can express our integral as

$$\int_{-1}^{+1} f(x) \, dx = \sum_{i=1}^{n} f(x_i)w_i \qquad (6.31)$$

with

$$w_i = 2\{\mathbf{P}^{-1}\}_{i0} \qquad (6.32)$$

The $w_i$ is the weight of $f(x_i)$ at the zeros of $P_n(x)$ and can be calculated. However, these values are also tabulated, for example by Abramowitz and Stegun in the *Handbook of Mathematical Functions* [6]. The following table lists the values for $n = 4$ and $n = 5$.

| Order | Abscissa $x_i$ | Weight $w_i$ |
|-------|----------------|--------------|
| $n = 4$ | $\pm 0.339981043584856$ | $0.652145154862546$ |
|         | $\pm 0.861136311594053$ | $0.347854845137454$ |
| $n = 5$ | $\pm 0.000000000000000$ | $0.568888888888889$ |
|         | $\pm 0.538469310105683$ | $0.478628670499366$ |
|         | $\pm 0.906179845938664$ | $0.236926885056189$ |

## Gaussian integration with Laguerre polynomials

The drawback of the Gauss–Legendre integration is that the limits have to be finite, because the Legendre polynomials are only defined on $[-1, 1]$. However, many times you will encounter integrals of the form:

$$\int_0^\infty f(x)\, dx \tag{6.33}$$

or

$$\int_{-\infty}^\infty f(x)\, dx \tag{6.34}$$

One type of integral commonly encountered in physics is of the type:

$$\int_0^\infty e^{-x} f(x)\, dx \tag{6.35}$$

This integral can be calculated using Gauss–Laguerre quadrature, where we are using Laguerre polynomials, which are orthogonal polynomials in the region $[0, \infty]$.

$$\int_0^\infty e^{-x} f(x)\, dx = \sum_{k=1}^N f(x_k) W_k \tag{6.36}$$

where the $x_k$ are the zeros of the Laguerre polynomials and the $W_k$ are the corresponding weights and are tabulated in [6]. Another way to look at the integration in $[0, \infty]$ is

$$\int_0^\infty f(x)\, dx = \int_0^\infty e^x e^{-x} f(x)\, dx$$

$$= \sum_{k=1}^N w(x_k)\, e^{x_k}\, f(x_k) \tag{6.37}$$

which is also tabulated in Abramowitz and Stegun [6]. You can also find these values at:

http://www.efunda.com/math/num_integration/num_int_gauss.cfm.

Many integrals involving Gaussian distributions are solvable by using Hermitian polynomials $H(x)$, which are defined in $[-\infty, \infty]$

$$\int_{-\infty}^\infty f(x)\, dx \approx \sum_{k=1}^N w_k f(x_k) \tag{6.38}$$

and integrals of the form

$$\int_{-1}^{1} \frac{f(x)\,dx}{\sqrt{1-x^2}} \tag{6.39}$$

can be solved with Chebyshev polynomials:

$$\int_{-1}^{1} \frac{f(x)\,dx}{\sqrt{1-x^2}} \approx \sum_{k=1}^{N} w_k f(x_k) \tag{6.40}$$

where in both cases the $x_k$ are the roots of the respective polynomials.

## 6.4 Exercises

1. Using your sin program, write a new program which integrates

$$f(x) = \int_{0}^{\pi} \sin(x)\,dx$$

   with $N$ intervals, where $N = 4,\ 8,\ 16,\ 256$ and $1024$ and compare the result for the trapezoid and Simpson methods.
2. Write a general use function or class in which you can give the function and integration limits.
3. Use your created function to solve the problem of a projectile with air resistance to determine the horizontal and vertical distances as well as the corresponding velocities as a function of time. This is a problem which you have to outline carefully before you attack it.
4. Use Laguerre integration to calculate the Stefan–Boltzmann constant.

# Chapter 7
# Solution of nonlinear equations

From school we know how to find the roots of a quadratic equation $ax^2 + bx + c = 0$, namely the solution $x = (-b \pm \sqrt{b^2 - 4ac})/2a$. The problem of finding the roots of a quadratic equation is a particular case of nonlinear equations $f(x) = 0$. The function $f(x)$ can be a polynomial, transcendental, or a combination of different functions, like $f(x) = \exp(x)\ln(x) - \cos(x) = 0$. In science we encounter many forms of nonlinear equations besides the quadratic ones. As a rule, it is difficult or not feasible to find analytic solutions. A few lines of computer code can find numerical solutions instantly. You may already have some experience with solving nonlinear equations with a programmable graphical calculator. In this chapter we consider a few simple but powerful methods for solving nonlinear equations $f(x) = 0$. However, the simplicity may be deceptive. Without proper understanding, even simple methods may lead you to poor or wrong solutions.

In the following, let $f(x)$ be a function of a single real variable $x$, and we will look for a real root on an interval $[a, b]$.

## 7.1 Bisection method

The bisection method is the simplest but most robust method. This method never fails. Let $f(x)$ be a continuous function on $[a, b]$ that changes sign between $x = a$ and $x = b$, i.e. $f(a)f(b) < 0$. In this case there is at least one real root on the interval $[a, b]$. The bisectional procedure begins with dividing the initial interval $[a, b]$ into two equal intervals with the middle point at $x_1 = (a + b)/2$. There are three possible cases for the product of $f(a)f(x_1)$

$$f(a)f(x_1) \begin{cases} < 0 & \text{there is a root in } [a, x_1] \\ > 0 & \text{there is a root in } [x_1, b] \\ = 0 & \text{then } x_1 \text{ is the root} \end{cases}$$

We can repeat the bisectional procedure for a new interval where the function $f(x)$ has a root. On each bisectional step we reduce by two the interval where the solution occurs. After $n$ steps the original interval $[a, b]$ will be reduced to the interval $(b-a)/2^n$. The bisectional procedure is repeated until $(b-a)/2^n$ is less than the given tolerance.

## 7.2 Newton's method

Newton's method exploits the derivatives $f'(x)$ of the function $f(x)$ to accelerate convergence for solving $f(x) = 0$ with the required tolerance. A continuous function $f(x)$ around the point $x$ may be expanded in a Taylor series:

$$f(x) = f(x_0) + (x - x_0)f'(x_0) + (x - x_0)^2 \frac{f''(x_0)}{2!} + (x - x_0)^3 \frac{f'''(x_0)}{3!} + \cdots \quad (7.1)$$

Suppose $x$ is the solution for $f(x) = 0$. If we keep the first two terms in our Taylor series, we obtain:

$$f(x) = 0 = f(x_0) + (x - x_0)f'(x_0) \quad (7.2)$$

Whence it follows that:

$$x = x_0 - \frac{f(x_0)}{f'(x_0)} \quad (7.3)$$

Because equation (7.2) corresponds to a linear approximation over $(x - x_0)$, we need more than one iteration to find the root. And the next iteration is:

$$x_{k+1} = x_k - \frac{f(x_k)}{f'(x_k)} \quad (7.4)$$

Usually the procedure requires only a few iterations to obtain the given tolerance. However, the method has three weak points: (i) a very slow approach to the solution when $f'(x) \to 0$ around the root; (ii) difficulty with local minima, leading to the next iteration value $x_{k+1}$ being far away; (iii) lack of convergence for asymmetric functions $f(a + x) = -f(a - x)$. We recommend using Newton's method when computer time is an issue and you know that the method will work. Otherwise, the bisection method will be the safer choice.

## 7.3 Method of secants

The secant method is a variation of Newton's method when the evaluation of derivatives is difficult. The derivative $f'$ of the continuum function $f(x)$ at point $x_0$ can be presented by

$$f'(x_0) = \frac{f(x_0) - f(x_1)}{x_0 - x_1} \quad (7.5)$$

and equation (7.4) becomes

$$x_{k+1} = x_k - \frac{(x_k - x_{k-1})f(x_k)}{f_k - f_{k-1}} \tag{7.6}$$

Therefore, one has to select two initial points to start the iterative procedure. The convergence of the secant method is somewhere between Newton's method and the bisection method.

## 7.4 Brute force method

All the methods above are designed to find a single root on an interval $[a,b]$. However, the following situations are possible: (i) there are two or more roots on the interval $[a,b]$; (ii) there is one root, but the function does not change sign, as in the equation $x^2 - 2x + 1 = 0$; (iii) there are no roots at all; (iv) there is a singularity, so that the function does change sign but the equation does not have a root. The bisection method in the last case will converge on the singularity if the interval has no roots.

The brute force method is a good approach for dealing with multiple roots. You split the original interval $[a,b]$ into smaller intervals with a stepsize $h$, applying some of the methods mentioned above to each subinterval. Care should be taken in selecting the size of $h$. If the interval is too large we may miss multiple zeros. On the other hand, choosing too small a step may result in time consuming calculations. A graphical analysis of the equation may help determine the most reasonable size for $h$.

## 7.5 Exercises

1. Write a program that implements the bisection method to find roots on the interval $[a,b]$.
2. Apply the program developed above to find a root between $x = 0$ and $x = 4$ for $f(x) = \exp(x)\ln(x) - \cos(x) = 0$.
3. Develop a program that can solve a nonlinear equation with Newton's method.
4. Compare the effectiveness of the bisection method and Newton's method for the equation $x^3 - 2x - 2 = 0$ which has a single root between $x = -4$ and $x = 2$.

5. Find the real zero of $x^2 - 2x + 1 = 0$ on $[-5, +5]$.

6. Write a program that implements the brute force method together with the bisection method for subintervals to solve $f(x) = \cos(2x) - 0.4x = 0$ on $[-4.0, +6.5]$.

# Chapter 8
# Differential equations

## 8.1 Introduction

Some of the most fundamental and frequently occurring mathematical problems in physics use differential equations. Whereas "regular" equations have numbers as a solution, the solution to a differential equation is a function which satisfies the equation. Because this is such an important topic, we will discuss several approaches to solving these equations.

Probably the first differential equation you encountered is Newton's second law:

$$\mathbf{F} = \frac{\mathrm{d}}{\mathrm{d}t}(m\mathbf{v}) = m\frac{\mathrm{d}\mathbf{v}}{\mathrm{d}t} = m\ddot{\mathbf{r}} \tag{8.1}$$

This is a second order differential equation (because you have a second derivative), which you can integrate provided the force $\mathbf{F}$ and the initial conditions are known. Two other common differential equations are the Schrödinger equation and the equation describing a harmonic oscillator.

Generally, differential equations are equations in which there is a function, derivatives of the function, and independent variables:

$$y' + 2xy = 0 \tag{8.2}$$

In this case $y$ is the function you are trying to find, $y'$ is its first derivative, and $x$ is the independent variable. One solution is $y = e^{-x^2}$ which you can test by substituting into equation (8.2).

## 8.2 A brush up on differential equations

Before we continue, here is a quick tour of the various equations called differential equations. There are two different classes: the ordinary differential equations (ODE) which contain functions of only one independent variable, and partial differential equations (PDE) having functions of several independent variables. An example of the latter is the time dependent Schrödinger equation, where the wave function $\Psi(\mathbf{r}, t)$ is a function of the space coordinates $\mathbf{r}$ and the time $t$. In this book we will limit ourselves to ordinary differential equations.

ODEs can themselves be grouped into subcategories according to their order $n$ and whether they are linear. In a linear equation, the function and its derivatives appear only in their first power and are not multiplied with each other. To illustrate this point the following equation

$$x^4 y''' = y \tag{8.3}$$

is linear while

$$x^4 y''' y' = y \tag{8.4}$$

and

$$x^4 y''' = y^2 \tag{8.5}$$

are not, because they contain the products $y''' y'$ and $y^2$ respectively.

However, all three equations are third order, because the highest derivative appearing is $y'''$.

The most general form for an $n$th order ordinary differential equation is:

$$F(x, y, y', y'', \dots, y^{(n)}) = 0 \tag{8.6}$$

When solving differential equations one has the further distinction between the general and a particular solution. We illustrate this with another example:

$$y' = y \tag{8.7}$$

The general solution of equation (8.7) is:

$$y = C\, e^x \tag{8.8}$$

where $C$ is an arbitrary constant, while $y = 2e^x$ or $y = e^x$ would each be a particular solution of equation (8.7). Finding the general solution is also called the "integration" of the equation.

The last two items to discuss are the distinction between homogenous and nonhomogeneous equations. In homogenous equations, each term contains either the function or its derivatives, but no other function of the independent variable. A good example of this kind is the simple harmonic oscillator:

$$m\frac{d^2x(t)}{dt^2} + kx(t) = 0 \qquad (8.9)$$

This is a linear (no higher power of $x(t), x''(t)$ or multiplications), second order homogeneous differential equation, well known from introductory physics. Also note that we now have $x$ as a function of $t$. However, adding an external force $F(t)$, as in the driven harmonic oscillator, leads to a non-homogeneous equation:

$$m\frac{d^2x(t)}{dt^2} + kx(t) = F_0 \cos \omega t \qquad (8.10)$$

In physics we frequently use second order differential equations. Determining what the initial conditions are is the first step in solving a problem. The reason for this is that every $n$th order ODE has $n$ integrating constants, which have to be known in order to solve the system. These constants can be either the initial values, as mentioned above, or boundary conditions, which will be determined at the integration limits. Examples are the initial position and velocity of a particle subject to a known force, or the boundaries of the potential well in the Schrödinger equation, respectively.

## 8.3 Introduction to the simple and modified Euler methods

Let us start with the simplest approach to solving an ODE, namely the Euler method. This is more for instructional purposes than any serious attempt to solve ODEs, especially in the case of the simple method. However, it is a good introduction to the chapter, because approaches like the Runge–Kutta formalism are really just sophisticated adaptations of the Euler method. As a starting point we look at the approximation for the derivative:

$$y'(x_0) \approx \frac{y(x_0 + h) - y(x_0)}{(x_0 + h) - (x_0)} \qquad (8.11)$$

We of course are interested in solving for $y(x_0 + h)$, requiring us to rewrite the equation as

$$y(x_0 + h) \approx y(x_0) + hy'(x_0) \qquad (8.12)$$

We have just calculated the next point $y(x_0 + h)$ using the previous point $y(x_0)$. In other words, if we know the derivative at $x_0$ and the step size $h$, we predict the value of $y$ at $x_0 + h$. You might also recognize this as the first two terms of the Taylor series. Clearly, only keeping the first two terms will not give us a very precise solution. You can guess from this that error propagation can become a problem.

## 8.4 The simple Euler method

As an example we will begin our discussion with a first order ODE

$$xy' + y = 0 \tag{8.13}$$

which is solvable analytically. If we rewrite this equation as

$$x\frac{dy}{dx} = -y \tag{8.14}$$

the following steps become quite clear:

$$\frac{dy}{y} = -\frac{dx}{x} \tag{8.15}$$

$$\int \frac{dy}{y} = -\int \frac{dx}{x} \tag{8.16}$$

$$\ln y = -\ln x + C \tag{8.17}$$

which results in

$$y = \frac{C}{x} \tag{8.18}$$

In Figure 8.1 the analytical solution is drawn.

We will now discuss how to solve this equation with numerical tools, starting with the simple Euler method. Using equation (8.12) and the fact that

$$y' = -\frac{y}{x} \tag{8.19}$$

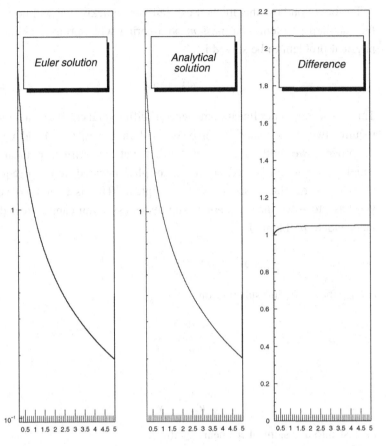

**Figure 8.1** Simple Euler, analytical solution and ratio of Euler over the analytical solution for a step size of 0.01.

we arrive at

$$y(x_0 + h) = y(x_0) - h\frac{y(x_0)}{x_0} \tag{8.20}$$

In this case we start from a point $x_0$, divide the interval between $x_0$ and $x_f$ into $N$ equidistant steps of size $h$ and calculate $y$ at $x_0, x_0 + h, x_0 + 2h, \ldots, x_0 + Nh$.

As you can see from Figure 8.1, the simple Euler approach does give a satisfactory result (depending on your standards); the difference between the numerical and analytical solutions is just a few percent. Part of the reason for the difference is that the step size in the program was rather coarse, only 0.01. If you choose a smaller step size, the Euler solution will more closely approximate the analytical result. However, the finer you make the step size, the more computing time you will need.

To show you how this method can fail very quickly, we will now discuss the standard problem of a mass $m$ on a spring with spring constant $k$. The physical problem to be solved is:

$$m\ddot{x} = -kx \qquad (8.21)$$

This is a second order linear homogeneous differential equation, and therefore requires two initial values to solve. We will choose $v_0(t=0)$ and $x_0(t=0)$. To proceed we will make use of the fact that any $n$th order linear differential equation can be reduced to $n$ **coupled** linear differential equations. (Notice the boldface used to write "coupled." This is to remind you that you have to solve these equations simultaneously; you cannot treat them as independent problems.)

$$y^{(n)} = f(x, y, y', y'', \ldots, y^{(n-1)}) \qquad (8.22)$$

can by the following substitution

$$u_1 = y'$$

$$u_2 = y''$$

$$\vdots$$

$$u_{n-1} = y^{(n-1)}$$

be cast into a system of $n$ linear equations

$$y' = u_1$$

$$u_1' = u_2$$

$$u_2' = u_3$$

$$u_{n-2}' = u_{n-1}$$

$$u_{n-1}' = f(x, y, u_1, u_2, \ldots, u_{n-1})$$

Now back to our problem of the harmonic oscillator. Because of the second derivative we will then have two equations, one for

$$u_1 = \frac{dx}{dt}$$

and one for

$$u_2 = \frac{dv}{dt} = \frac{d^2x}{dt^2}$$

which are now both linear. Writing this for the harmonic oscillator we end up with:

$$m\frac{dx}{dt} = mv \tag{8.23}$$

$$m\frac{dv}{dt} = -kx \tag{8.24}$$

In order to show that this is not just a mathematical shuffling around we have kept the $m$ in equations (8.23) and (8.24) to make the connection for the more advanced student to the Hamiltonian equations, which are:

$$m\frac{dx_i}{dt} = p_i \tag{8.25}$$

$$\frac{dp_i}{dt} = F_i \tag{8.26}$$

Again, we already know the result of this problem. If $v_0(t=0)=0$ and $x_0(t=0)=A$ we have:

$$x(t) = A\cos(\omega t) \tag{8.27}$$

$$v(t) = -\omega A\sin(\omega t) \tag{8.28}$$

$$\omega = \sqrt{\frac{k}{m}} \tag{8.29}$$

We will use conservation of energy to check our numerical solution later. In the next step we have set up the two equations as they appear for the numerical solution with $m=k=1$. Using equations (8.12), (8.23) and (8.24) and writing $\Delta t$ for $h$, the first one becomes

$$x(t+\Delta t) = x(t) + \Delta t v(t) \tag{8.30}$$

while the second is

$$v(t+\Delta t) = v(t) - \Delta t x(t) \tag{8.31}$$

In Appendix D you will find the source code for the program.

As you can see in Figure 8.2, this simple approach fails miserably. Even though we get the expected oscillating result, a closer look reveals that the

**Figure 8.2** Harmonic oscillator solution using the simple Euler model.

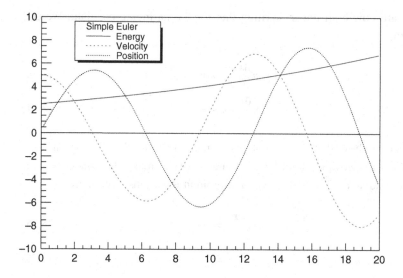

amplitudes are growing with time, violating conservation of energy. This would be a perfect perpetuum mobile, with the energy always increasing. This failure is caused by our truncation of the Taylor series.

## 8.5 The modified Euler method

We can improve our results by being slightly more sophisticated. In the simple Euler method we used the derivative calculated at the beginning of the interval (i.e., at $x$) to predict the value of $y$ at $x + h$. The modified Euler method estimates the solution by using the derivative at the midpoint between $x$ and $x + h$, similar to taking the central difference described in Chapter 5. Again we start the discussion with the first problem encountered, equation (8.13)

$$xy' + y = 0$$

$$y(x_0 + h) = y(x_0) + hy'(x_0 + h/2) \tag{8.32}$$

We will first determine $y$ at the midpoint by writing

$$y\left(x_0 + \frac{h}{2}\right) = y(x_0) + \frac{h}{2}y'(x_0) \tag{8.33}$$

and then insert this value into our equation for $y'$, effectively propagating this now over the whole interval:

$$y(x_0 + h) = y(x_0) + h\frac{y(x_0 + h/2)}{x_0 + h/2} \tag{8.34}$$

In Figure 8.3 the results from the halfpoint method are shown. As you can see the agreement is now excellent.

Let us now apply this improved method to the harmonic oscillator. First we calculate the values for $x$ and $v$ at the midpoint of the interval using the simple Euler method:

$$v_{\mathrm{mid}} = v\left(t + \frac{\Delta t}{2}\right) = v(t) - \frac{\Delta t}{2}x(t) \tag{8.35}$$

$$x_{\mathrm{mid}} = x\left(t + \frac{\Delta t}{2}\right) = x(t) + \frac{\Delta t}{2}v(t) \tag{8.36}$$

Using the velocity and position at these points we can then propagate our system forward to $t + \Delta t$:

$$v(t + \Delta t) = v(t) - \Delta t x_{\mathrm{mid}} \tag{8.37}$$

$$x(t + \Delta t) = x(t) + \Delta t v_{\mathrm{mid}} \tag{8.38}$$

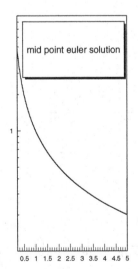
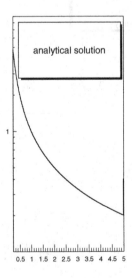
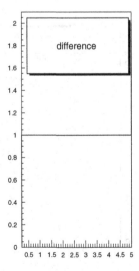

**Figure 8.3** Modified Euler method, analytical solution and ratio of Euler over the analytical solution for a step size of 0.01.

To implement this in your code, use your previous code with the simple Euler method and add two lines for the midpoint calculation. Replace

```
v_new[loop]=v_old-step*x_old; // calculate the forward
velocity
X_new[loop]=x_old+step*v_old; // calculate the forward
position
```

with the following new code segment:

```
V_mid=v_old-step/2.*x_old; // calculate the midpoint
\index{midpoint} velocity
x_mid=x_old+step/2.*v_old; // calculate the midpoint
\index{midpoint} position
v_new[loop]=v_old-step*x_mid; // calculate the forward
velocity
x_new[loop]=x_old+step*v_mid; // calculate the forward
position
```

In Figure 8.4 you can see the result of the new method. Neither velocity nor position diverge, and energy is now conserved. This is much better but still not perfect. A closer look reveals that for either a large step size or a loop over many periods, the energy still diverges (Figure 8.5).

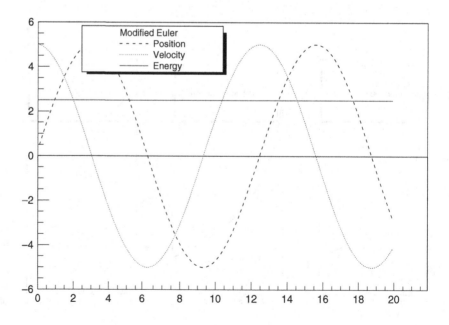

**Figure 8.4** Harmonic oscillator with the modified Euler model using two periods.

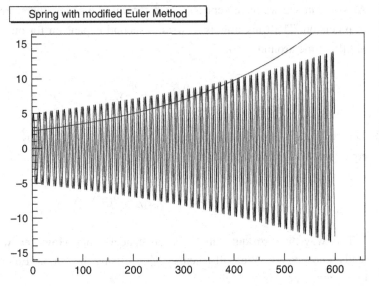

**Figure 8.5** Harmonic oscillator with the modified Euler model. The step size has been chosen to be 0.3 instead of 0.1 and the number of periods is now 50. Note how the amplitude increases and the energy diverges.

## 8.6 Runge–Kutta method

It is now time to introduce the most widely used method to solve differential equations numerically: the fourth order Runge–Kutta (RK) method. This is an improvement of the modified Euler method, which is itself a second order Runge–Kutta method. In the following we will use a slightly more formalistic approach. To understand this better, we start with the Taylor series for a function (where we write $y_n$ for $y(x_n)$ and $x_n = x + nh$)

$$y_{n+1} \cong y(x_n + h) = y_n + hy'_n + \frac{h^2}{2}y''_n + \mathcal{O}(h^3) \tag{8.39}$$

Just keeping the first two terms, we have the simple Euler method which has an error of order $h^2$. Writing

$$y'_n = f(x_n, y_n) \tag{8.40}$$

i.e., the derivative of $y$ at $x_n$. We can express the simple method as:

$$y_{n+1} = y_n + hf(x_n, y_n) \tag{8.41}$$

This is the expression for the simple method, which as you can see has an error of order $h^2$. The next step is to get to the midpoint method. In order to do this we now assume we know $y_{n+1/2}$ and expand this in a Taylor series around $x - h/2$, giving us $y_n$:

$$y_n = y_{n+1/2} - \frac{h}{2}y'_{n+1/2} + \frac{h^2}{2}y''_{n+1/2} + \mathcal{O}(h^3) \tag{8.42}$$

As you can see we have kept one more term in the series, therefore gaining in precision. The trick now is to write a second expansion for our function, but this time around $x + h/2$:

$$y_{n+1} = y_{n+1/2} + \frac{h}{2}y'_{n+1/2} + \frac{h^2}{2}y''_{n+1/2} + \mathcal{O}(h^3) \qquad (8.43)$$

This will then allow us to get rid of the unknown $y''_{n+1/2}$ by subtracting equation (8.42) from (8.43) which results in:

$$y_{n+1} = y_n + hy'_{n+1/2} + \mathcal{O}(h^3) \qquad (8.44)$$

This way the term quadratic in $h$ has canceled out. However, we do not really know $y_{n+1/2}$. We will calculate $y'_{n+1/2}$ using the simple Euler method to approximate this midpoint:

$$y'_{n+1/2} = y_n + \frac{h}{2}y'_n \qquad (8.45)$$
$$= f\left(x_{n+1/2}, y_n + \frac{h}{2}f(x_n, y_n)\right)$$
$$= f(x_{n+1/2}, y_{n+1/2})$$

Now we can use this result to start writing down an iterative algorithm, which requires you to use the known derivative at $x_n$ (namely $k_1$), to calculate from this the midpoint derivative at $k_2$

$$k_1 = f(x_n, y_n) \qquad (8.46)$$
$$k_2 = f\left(x_n + \frac{h}{2}, y_n + \frac{hk_1}{2}\right) \qquad (8.47)$$

and finally propagate this through the whole interval:

$$y_{n+1} = y_n + hk_2 \qquad (8.48)$$

We have just re-derived the modified Euler method, but now in terms of what is called a second order Runge–Kutta method (the method is called $n$th order if its error is $\mathcal{O}(h^{n+1})$). If we iterate these steps a few more times, we can cancel higher and higher orders of $h$. In practice people usually do this up to fourth or fifth order, which is shown in the next equations (fourth order):

$$k_1 = f(x_n, y_n) \tag{8.49}$$

$$k_2 = f\left(x_n + \frac{h}{2}, y_n + \frac{hk_1}{2}\right) \tag{8.50}$$

$$k_3 = f\left(x_n + \frac{h}{2}, y_n + \frac{hk_2}{2}\right) \tag{8.51}$$

$$k_4 = f(x_n + h, y_n + hk_3) \tag{8.52}$$

with

$$y_{n+1} = y_n + h\left(\frac{k_1}{6} + \frac{k_2}{3} + \frac{k_3}{3} + \frac{k_4}{6}\right) + \mathcal{O}(h^5) \tag{8.53}$$

This is the most straightforward Runge–Kutta calculation and very useful for quick integrations of ODEs, especially if the function is smooth and slowly varying, and if computer time is not an issue.

Again, we will use our harmonic oscillator as an example hopefully to shed more light on this method. We start with our two equations, replacing our second order equation with the two linear equations (8.23) and (8.24). In the following, for instructional purposes, we have chosen a slightly more cumbersome way to program the task. Later, in the adaptive step control program, you will find the elegant solution with function calls for determining the derivatives.

The first thing we have to do is determine the derivatives $f(t_n, x_n)$ and $f(t_n, v_n)$ for $x$ and $v$ ($x$ is now the dependent variable and $t$ is the independent one). We have to find $(x(t), v(t))$. From $dx/dt = v$ we get

$$f(t_n, x_n) = v_n \tag{8.54}$$

$$k_{x1} = f(t_n, x_n) \tag{8.55}$$

$$x_{\text{temp}} = x_n + \frac{h}{2}k_{x1} \tag{8.56}$$

and similarly for $v$

$$f(t_n, v_n) = -x_n \tag{8.57}$$

$$k_{v1} = f(t_n, v_n) \tag{8.58}$$

$$v_{\text{temp}} = v_n + \frac{h}{2}k_{v1} \tag{8.59}$$

Having now found our first two derivatives $k_{x1}$ and $k_{v1}$, we need to find the next pair $k_{x2}$ and $k_{v2}$ for $t + h/2$. The temporary values $x_{\text{temp}}$ and $v_{\text{temp}}$ are the values of the derivatives at this point, so we write:

$$f\left(t_n + \frac{h}{2}, x_n + \frac{k_{x1}}{2}\right) = v_{\text{temp}} \tag{8.60}$$

$$k_{x2} = f\left(t_n + \frac{h}{2}, x_n + \frac{k_{x1}}{2}\right) \tag{8.61}$$

$$x_{\text{temp}} = x_n + \frac{h}{2}k_{x2} \tag{8.62}$$

and similarly for $v$

$$f\left(t_n + \frac{h}{2}, v_n + \frac{k_{v1}}{2}\right) = -x_{\text{temp}} \tag{8.63}$$

$$k_{v2} = f\left(t_n + \frac{h}{2}, v_n + \frac{k_{v1}}{2}\right) \tag{8.64}$$

$$v_{\text{temp}} = v_n + \frac{h}{2}k_{v2} \tag{8.65}$$

The third step then becomes:

$$f\left(t_n + \frac{h}{2}, x_n + \frac{k_{x2}}{2}\right) = v_{\text{temp}} \tag{8.66}$$

$$k_{x3} = f\left(t_n + \frac{h}{2}, x_n + \frac{k_{x2}}{2}\right) \tag{8.67}$$

$$x_{\text{temp}} = x_n + hk_{x3} \tag{8.68}$$

and similarly for $v$

$$f\left(t_n + \frac{h}{2}, v_n + \frac{k_{v2}}{2}\right) = -x_{\text{temp}} \tag{8.69}$$

$$k_{v3} = f\left(t_n + \frac{h}{2}, v_n + \frac{k_{v2}}{2}\right) \tag{8.70}$$

$$v_{\text{temp}} = v_n + hk_{v3} \tag{8.71}$$

And finally the last two derivatives are just

$$k_{x4} = v_{\text{temp}} \tag{8.72}$$

$$k_{v4} = -x_{\text{temp}} \tag{8.73}$$

The complete solution for both variables $x$ and $v$ becomes:

$$x(t+h) = x(t) + \frac{h}{6}(k_{x1} + 2k_{x2} + 2k_{x3} + k_{x4}) \tag{8.74}$$

$$v(t+h) = v(t) + \frac{h}{6}(k_{v1} + 2k_{v2} + 2k_{v3} + 2k_{v4}) \tag{8.75}$$

The new values at $t+h$ now become the starting values for the next loop and you go back to the top. The following is a little code snippet for this problem.

```
while (time <100.)
{
kx1 = v_old;
 x_temp = x_old+step *kx1/2.;
kv1 = -x_old;
 v_temp = v_old+step*kv1/2.;

kx2 = v_temp;
 x_temp = x_old+step/2*kx2;
  kv2 = -x_temp;
 v_temp = v_old+step/2*kv2;

kx3 = v_temp;
 x_temp = x_old+step*kx3;
  kv3 = -x_temp;
 v_temp = v_old+step*kv3;

kx4 = v_temp;
kv4 = -x_temp;

x_new = x_old+1./6.*(kx1+2*kx2+2*kx3+kx4)*step;
v_new=v_old+1./6.*(kv1+2*kv2+2*kv3+kv4)*step;

time=time+step;

v_old=v_new;
x_old=x_new;
  loop=++loop;
}
```

One of the remaining tasks is to estimate the quality of your answer. You could run your code with two different step sizes and see whether the difference is within acceptable limits. If you find that this is not the case, usually you would make your step size smaller and try again, as we did in the modified Euler method. However, when the function is not slowly varying, then it will be more productive to optimize your algorithm. The methods so far described used the same step size over the full interval, an approach which can be potentially wasteful in computer time. A region where the function is rapidly varying needs a much smaller step size than a region where the function is slowly varying. One way out of this predicament is to use an algorithm which has a step size that is automatically adjusted for different regions.

## 8.7 Adaptive step size Runge–Kutta

The traditional and also most intuitive way to vary the step size adaptively is the step doubling technique. The procedure involves first calculating the next point, $y_1$, with a step size of $h$ and then a second point, $y_2$, with step size $h/2$. You then take the two calculated values $y_1$ and $y_2$ and compare the difference $\Delta = y_2 - y_1$ with a predefined acceptable error $\epsilon$. If $\Delta$ is less than $\epsilon$, we can use step size $h$, otherwise we would repeat this with smaller step sizes $h/2$, $h/4$, until we find a satisfactory $\Delta$. However, this method requires calculating the derivatives at both $f(x+h, y+h)$ and $f(x+h/2, y+h/2)$, a rather time consuming process.

An improvement to this approach was introduced by Fehlberg [7], who found a fifth order Runge–Kutta method, using six function evaluations. His improvement came from the fact that if he took a different combination of the six function evaluations, he got an additional fourth order method. Comparing the two solutions by requiring that their difference is smaller than a given value $\epsilon$ you will get a measure of the error resulting from the truncation. If this difference is larger than desired, the program reduces the step size automatically and recalculates the function values at the new points in $x$.

In the following, we outline how to make an estimate for the step size. For an $n$th order Runge–Kutta solution $y^n$,[1] the deviation from the exact result is of the order $\mathcal{O}(h^{n+1})$, and similarly for an $(n+1)$th order solution $y^{n+1}$, we get $\mathcal{O}(h^{n+2})$. If we take the difference between the two solutions

$$y^n - y^{n+1} = \Delta \approx h^{n+1} < \epsilon \tag{8.76}$$

[1] In this context $y^n$ and $y^{n+1}$ represent an $n$th and $(n+1)$th order solution and not the $n$th or $(n+1)$th derivative.

for small $h$. If our difference is smaller than $\epsilon$ we want to use a larger step size for the next interval, one which would produce a $\Delta_{\text{new}} = \epsilon$. This leads us to an estimate for a new step size $h_{\text{new}}$ from

$$\left| \frac{\Delta_{\text{new}}}{\Delta_{\text{current}}} \right| = \left| \frac{\epsilon}{\Delta_{\text{current}}} \right| \approx \left( \frac{h_{\text{new}}}{h_{\text{current}}} \right)^{1/n} \tag{8.77}$$

This leads to

$$h_{\text{new}} = h_{\text{current}} \left| \frac{\epsilon}{\Delta_{\text{current}}} \right|^{1/n} \tag{8.78}$$

If the difference is larger than $\epsilon$, calculated from equation (8.78), $h_{\text{new}}$ will be smaller than $h_{\text{current}}$ and we would go back and use the new $h_{\text{new}}$ as the step size. However, you have to guard against the possibility that $h_{\text{new}}$ becomes smaller than the least significant figure in $x$. This would cause an infinite loop. To prevent this you have to put a check in your program. In addition, because we have derived this equation with a lot of "..approximately.." we should also put a safety factor in it, which in effect makes the reduction for $h_{\text{new}}$ not as large as equation (8.78) would allow. Commonly a factor of 0.9 is used, leading to:

$$h_{\text{new}} = 0.9 h_{\text{current}} \left| \frac{\epsilon}{\Delta_{\text{current}}} \right|^{1/n} \tag{8.79}$$

Today you can choose between many variations, but all follow the same recipe for a $k$th order Runge–Kutta:

$$k_1 = hf(x_n, y_n) \tag{8.80}$$

$$k_2 = hf(x_n + a_2 h, y_n + b_{21} k_1) \tag{8.81}$$

$$k_{k+1} = hf(x_n + a_k h, y_n + \Sigma_{j=1}^{k} b_{k+1j} k_n) \tag{8.82}$$

From this you can then calculate the $k$th and $(k-1)$th order with

$$y_{n+1} = y_n + \sum_{j=1}^{k+1} c_j k_j \qquad k\text{th order} \tag{8.83}$$

$$y_{n+1} = y_n + \sum_{j=1}^{k+1} c_j^* k_j \qquad (k-1)\text{th order} \tag{8.84}$$

the difference being in the $c_j$ and $c_j^*$. Below is a table for fourth and fifth order solutions [8].

| $i$ | $a_i$ | $b_{ij}$ | | | | | $c_i$ | $c_i^*$ |
|---|---|---|---|---|---|---|---|---|
| 1 | | | | | | | $\frac{37}{378}$ | $\frac{2825}{27648}$ |
| 2 | $\frac{1}{5}$ | $\frac{1}{5}$ | | | | | 0 | 0 |
| 3 | $\frac{3}{10}$ | $\frac{3}{40}$ | $\frac{9}{40}$ | | | | $\frac{250}{621}$ | $\frac{18575}{48384}$ |
| 4 | $\frac{3}{5}$ | $\frac{3}{10}$ | $-\frac{9}{10}$ | $\frac{6}{5}$ | | | $\frac{125}{594}$ | $\frac{13525}{55296}$ |
| 5 | 1 | $-\frac{11}{54}$ | $\frac{5}{2}$ | $-\frac{70}{27}$ | $-\frac{35}{27}$ | | 0 | $\frac{277}{14336}$ |
| 6 | $\frac{7}{8}$ | $\frac{1631}{55296}$ | $\frac{175}{512}$ | $\frac{575}{13824}$ | $\frac{44275}{110592}$ | $\frac{253}{4096}$ | $\frac{512}{1771}$ | $\frac{1}{4}$ |

## 8.8 The damped oscillator

In this section we will discuss the oscillator, but now we will also have a retarding force. We will use this problem to clarify the use of the Runge–Kutta method with adaptive step size control. As you know, the simple harmonic oscillator, also called a free oscillator, is an oversimplification of the real world. This ideal device, once started, would continue to run infinitely. In the real world, however, we always have energy dissipated because of frictional losses, and eventually the motion dies out. A reasonable assumption is that the frictional force resisting the motion of a body is some power of the velocity of this body, i.e.,

$$\mathbf{F} \propto -b|v^n|\hat{\mathbf{v}} \tag{8.85}$$

where the minus sign expresses the fact that the force is directed in the opposite direction to the movement. This leads to the following equation for a mass $m$ moving under the influence of a restoring force $-kx$ and a retarding force $-b\dot{x}$

$$m\ddot{x} + b\dot{x} + kx = 0 \tag{8.86}$$

where we have assumed that the retarding force is linear in the velocity. Rewriting this as

$$\ddot{x} + 2\beta\dot{x} + \omega_0^2 x = 0 \tag{8.87}$$

we express the problem in the more familiar form with $\beta = b/2m$ the damping parameter, and $\omega_0^2 = k/m$ the angular frequency of the system.

First we will solve this analytically. Differential equations of the type

$$y'' + ay' + by = 0 \tag{8.88}$$

can always be solved by making the Ansatz:

$$y = e^{rx} \tag{8.89}$$

$$y' = re^{rx} \qquad y'' = r^2 e^{rx} \tag{8.90}$$

Using this in equation (8.88) leads to the following algebraic equation:

$$r^2 + ra + b = 0 \tag{8.91}$$

with the solution

$$r = -\frac{a}{2} \pm \sqrt{\frac{a^2 - 4b}{4}} \tag{8.92}$$

If the two roots $r_1$ and $r_2$ are not identical, then

$$y = c_1 e^{r_1 x} + c_2 e^{r_2 x} \tag{8.93}$$

is the general solution. In the case $r_1 = r_2 = r$ it can be shown that $xe^{rx}$ is also a solution, and because $e^{rx}$ and $xe^{rx}$ are linearly independent (the Wronskian determinant does not vanish),

$$y = c_1 e^{rx} + c_2 x e^{rx} \tag{8.94}$$

will be the general solution. Now, going back to our damped harmonic oscillator, our general solution of equation (8.87) becomes:

$$x(t) = e^{-\beta t} \left[ A_1 e^{\sqrt{\beta^2 - \omega_0^2}\, t} + A_2 e^{-\sqrt{\beta^2 - \omega_0^2}\, t} \right] \tag{8.95}$$

## Application of the adaptive step size Runge–Kutta method

To demonstrate the use of the adaptive step size Runge–Kutta method, we use the damped oscillator and explain the functions **rk4_stepper** and **rk4_step** in more detail. Both routines are general use routines we have written and are not limited to second order ODEs. The user, in this case you, has to provide two programs: one program supplies the input (e.g., initial conditions, starting step size), and the second program supplies the two linear differential equations.

Because the damped oscillator is a second order ODE, we will need to separate this equation into two first order differential equations. This can be achieved by setting,

$$y[0] = x$$
$$y[1] = \frac{dy[0]}{dt} = \dot{x}$$
$$\frac{dy[1]}{dt} = \ddot{x}$$

where we now have to integrate two equations simultaneously at different locations in $t$. To set up our problem for the form used in the Runge–Kutta mechanism we use

$$f_0 = \dot{x} \tag{8.96}$$

$$f_1 = -\omega^2 x - 2\beta\dot{x} \tag{8.97}$$

This is the heart of the routine calculating the derivative at various points in time $t$ and will be called from rk4_step.

```
#include "damp.h"
void deri (int nODE\index{ODE}, double x, double y[ ],
double f[ ])
// calculate the derivatives\index{derivatives} for the damped
oscillator
{
f[0]=y[1];
f[1]=-omega2*y[0]-2*beta*y[1];
return;
}
```

Now we come to the central pieces of the code, rk4_stepper and rk4_step. The first is used to control the step size in $t$ and will change the size depending on the precision of the solution found by rk4_step. rk4_step propagates the solution forward by one time slice and returns the integrated solutions for both equations as well as the differences between the respective fifth and sixth solutions. As we have seen in equation (8.76), this is a measure of the deviation from the "true" result. The stepper routine will then determine whether we have achieved the given precision or whether we need to reduce the step size. If the step size has to be reduced, rk4_step is called again with the smaller step. However, there is the problem that we could burn up a lot of computing time by continually reducing our step size and trying again and again until we reach the desired precision. Therefore, the program checks how much the step size has been reduced and exits when it has become smaller than a factor of 10. This means that you either have to decrease your desired precision or make the starting step size smaller. If you choose too small a step size, the program will increase your step size, but only to a given limit. rk4_stepper also opens a file where the integrated values will be written away. Here is the function rk4_stepper with additional documentation, especially in reference to our damped oscillator problem:

```
void rk4_stepper(double y_old[ ], int nODE, double xstart,
double xmax, double hstep, double eps, int &nxstep)
```

The next section has to do with the declarations:

| | |
|---|---|
| y_old[ ] | array of starting values for y |
| nODE | the degree of the ODE, in our case 2 |
| xstart | starting value, here begin in time |
| xmax | end of time |
| hstep | the beginning step size |
| eps | chosen precision |
| nxstep | returns the number of steps taken |

```
void rk4_step(double *, double *, double *, int,double,double);

double heps; // the product of the step size and the chosen error
double yerrmax=.99; // the max error in a step,
int i=0;
double const REDUCE=-.22 ; // reduce stepsize power
double esmall; // the lower limit of precision, if the result is
smaller
 //than this we increase the step size
double ydiff[nODE];
double y[nODE];
double hnew; // new step size
double x0;
```

```
double xtemp; // temporary x for rk4\_step
double step_lolim; // lower limit for step size
double step_hilim; // upper limit for step size

// here we create a file for storing the calculated functions
 ofstream out_file;
 out_file.open("rk4.dat");
 out_file.setf(ios::showpoint | ios::scientific | ios::
 right);
 x0=xstart;
 for(i=0 ; i<=nODE-1 ; i++)
 {
  // store starting values
 out_file << x0 << " "<<y_old[i]<<"\n";
 }
 esmall=eps/50.;
 heps=eps*hstep;

 step_lolim=hstep/10.; // the step size should not go lower
 than 10%
 step_hilim=hstep*10.; // We don't want larger step size
  for (i=0 ; i<=nODE-1 ; i++)
  {
 y[i]=y_old[i];
  }
```

And here comes the loop, where we break out once the error is smaller than the preset precision. You can see that this is checked for every ODE to ensure that all integrals reach the required accuracy. If the step size is too large, it is reduced and the step is repeated.

```
for( ; ; )
 {
 yerrmax=.99;
  for( ; ; )
  {
 xtemp=x0+hstep;
 rk4_step(y,ydiff,y_old,nODE,hstep,xtemp);
  for(i=0;i<=nODE-1;i++)
  {
 yerrmax=max(yerrmax,heps/ydiff[i]);
  }
```

```
if(yerrmax > 1.) break;
 hstep=.9* hstep* pow(yerrmax,REDUCE);
heps=hstep* eps;

// error if step size gets too low
 if(hstep<step_lolim)
 {
cout << "rk4_stepper: lower step limit reached; try
lower starting"
 << "step size \n";
cout << "I will terminate now \n";
exit(0);
 }
  }
```

This section will call `rk4_step` until the planned precision has been achieved or will terminate if `hstep` becomes smaller than `step_lolimit`. Because this function integrates many ODEs, every integral has to be checked for precision:

```
yerrmax=max(yerrmax,heps/ydiff[i]);
```

And, finally, the last part of the routine where x is propagated by `hstep` and the data are stored in the file.

```
// go further by one step
// first check if our step size is too small
if (yerrmax>1./esmall)hstep=hstep* 2.;

// set upper limit for step size
if(hstep>step_hilim){
hstep=step_hilim;
}
for(i=0 ; i<=nODE-1 ; i++)
{
 y_old[i]=y[i];
// store data in file rk4.dat
out_file << xtemp << " "<<y[i]<<"\n";
}

//   x0 +=hstep;
x0=xtemp;
nxstep=nxstep+1;
if(x0>xmax)
```

```
{
cout << nxstep;
 out_file.close( ); // close data file
 return;
}
 }
 return;
```

This leaves us to discuss the routine which actually calculates the integral for one step, rk4_step:

```
void rk4_step(double * y,double * ydiff,double *y_old,int nODE,
double step, double x0)
```

| | |
|---|---|
| y | array of the calculated fifth order integral |
| ydiff | array of difference between fifth and sixth orders |
| y_old | array of starting values for this step |
| x0 | the current $x$ value for $y$, in our case the time |

The next block has to do with setting up the constants for the Runge–Kutta method, and defining the f-arrays, where the differential will be calculated.

```
void deri (int , double ,double *, double *);
// user supplied routine
// which calculates derivative
// setup the constants first
// the coefficients for the steps in x
double const a2=1./5. ,a3=3./10., a4=3./5., a5=1. ,a6=7./8. ;
// coefficients for the y values
double const b21=1./5. ;
double const b31=3./40. , b32=9./40. ;
double const b41=3./10. , b42=-9./10., b43=6./5. ;
double const b51=-11./54. , b52=5./2. , b53=-70./20. ,
b54=-35./27.;
double const b61=1631./55296. , b62=175./312. ,
b63=575./13824. ;
double const b64=44275./110592. , b65=253./1771.;//

coefficients for y(n-th order)
double const c1=37./378. , c2=0. , c3=250./621., c4=125./594.,
c5=0., c6=512./1771.;

// coefficients for y(n+1-th order)
double const cs1=2825./27648., cs2=0., cs3=18575./48384.,
```

```
cs4=13525./55296., cs5=277./14336., cs6 = 1./4. ;

// the calculated values f
double f[nODE] , f1[nODE] , f2[nODE] , f3[nODE], f4[nODE],
    f5[nODE];

// the x value
double x;
double yp[nODE];
int i;
```

In the last section, first the derivatives are calculated and then the equations corresponding to (8.84) shown above.

```
// here starts the calculation of the RK\index{RK} parameters
  deri (i,x,y,f);
for(i=0; i<=nODE-1;i++) // 1. step
{
 y[i]=y_old[i]+b21*step*f[i];
 }
 x=x0+ a2*step;

deri (i,x,y,f1);

for (i=0; i<=nODE-1;i++) //2. step
{
 y[i]=y_old[i]+b31*step*f[i]+b32*step*f1[i];
 }
x=x0+ a3*step;

deri (i,x,y,f2);
for(i=0; i<=nODE-1;i++) //3. step
{
 y[i]=y_old[i]+b41*step*f[i]+b42*step*f1[i]+b43*step*f2[i];
 }
x=x0+ a4*step;

deri (i,x,y,f3);
for (i=0; i<=nODE-1;i++) //4. step
{
 y[i]=y_old[i]+b51*step*f[i]+b52*step*f1[i]+b53*step*
 f2[i]+b54*step*f3[i];
```

```
}
x=x0+ a5*step;

  deri (i,x,y,f4);
for (i=0; i<=nODE-1;i++) //5. step
{
y[i]=y_old[i]+b61*step*f[i]+b62*step*f1[i]+b63*step*
f2[i]+b64* step*f3[i]+b65*step*f4
 }
x=x0+ a6* step;

  deri (i,x,y,f5);
for (i=0; i<=nODE-1;i++) //6. step
{
y[i]=y_old[i]+step*(c1*f[i]+c2*f1[i]+c3*f2[i]+c4*f3[i]+
c5*f4[i]+c6*f5[i]);
yp[i]=y_old[i]+step*(cs1*f[i]+cs2*f1[i]+cs3*f2[i]+cs4
*f3[i]+cs5*f4[i]+cs6*f5[i]); ydiff[i]=fabs (yp[i]-y[i]);
 }
 return;
}
```

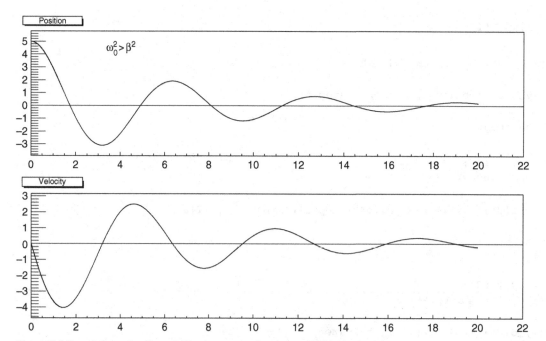

**Figure 8.6** Result from the Runge–Kutta program for the underdamped case.

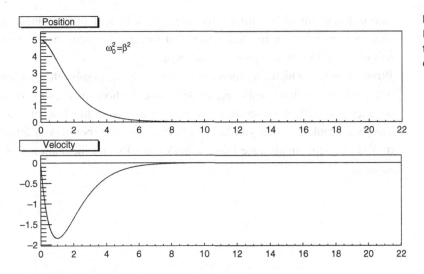

**Figure 8.7** Result from the Runge–Kutta program for the critically damped case.

One could have optimized this code by calculating yp[i] and taking the difference using y[i] and constants, where the difference has already been taken into account. However, for clarity and readability we prefer this method.

Now we are at the point were we can test our program and check whether the result is correct. In the first step we would run this code without any damping and check that the results for the position and velocity are indeed sinusoidal. We show here only the results for the cases of underdamped $\omega_0^2 > \beta^2$ (Figure 8.6) and critically damped $\omega_0^2 = \beta^2$ (Figure 8.7) oscillators.

## 8.9 Exercises

1. Write a program which uses the simple Euler method to solve for

$$y' = \frac{y}{x} - \frac{x^2}{y^2}$$

Use $y(1) = 1$ and step sizes of $h = 0.05, 0.1$ and $0.20$ for $0 < x < \sqrt[3]{e}$.

2. Modify your program for the same problem but using the modified Euler method.

3. Use the adaptive step size program to solve the two dimensional harmonic oscillator:

$$\mathbf{F} = -k\mathbf{r}$$

Use different initial conditions for $x(0)$ and $y(0)$ and plot $x$ versus $y$. Modify your program in such a way that the angular frequencies in $\omega_x$ and $\omega_y$ are different and plot $x$ versus $y$ again.

4. Write a program which uses fourth order Runge–Kutta to solve the problem of a projectile with air resistance to determine the horizontal and vertical distances as well as the corresponding velocities as a function of time. This is a problem which you have to outline carefully before you attack it. The program should take initial muzzle velocity and inclination angle as input.

# Chapter 9
# Matrices

Matrices are among the most important mathematical objects in physics. Two principal computational problems are associated with matrices: linear systems of equations and eigenvalue problems.

## 9.1 Linear systems of equations

A set of simultaneous linear equations can be written in the form

$$a_{11}x_1 + a_{12}x_2 + \cdots + a_{1n}x_n = b_1$$

$$a_{21}x_1 + a_{22}x_2 + \cdots + a_{2n}x_n = b_2$$

$$\cdots\cdots\cdots\cdots\cdots\cdots\cdots\cdots\cdots\cdots\cdots$$

$$a_{m1}x_1 + a_{m2}x_2 + \cdots + a_{mn}x_n = b_m$$

(9.1)

where $x_j$, $j = 1, \ldots, n$ is a set of unknowns, $b_i$, $i = 1, \ldots, m$ are the right-hand side coefficients, $a_{i,j}$ are the coefficients of the system. Three cases are possible: $m > n$, $m = n$, and $m < n$. If the number of equations $m$ is larger than the number of unknowns $n$, the system of equations is *overdetermined*. This case is quite common in data processing. When $m < n$ the system of equations is *underdetermined* and cannot be solved. In this section we will consider the case where $m = n$, the number of unknowns is equal to the number of equations. It is this case that corresponds to solving problems in physics from "first principles."

Using matrix notation the system (9.1) for the $n \times n$ case can be written as

$$
\begin{pmatrix}
a_{11} & a_{12} & \cdots & a_{1n} \\
a_{21} & a_{22} & \cdots & a_{2n} \\
\cdots & \cdots & \cdots & \cdots \\
a_{n1} & a_{n2} & \cdots & a_{nn}
\end{pmatrix}
\begin{pmatrix}
x_1 \\
x_2 \\
\cdots \\
x_n
\end{pmatrix}
=
\begin{pmatrix}
b_1 \\
b_2 \\
\cdots \\
b_n
\end{pmatrix}
$$

(9.2)

or

$$\mathbf{A} \cdot \mathbf{x} = \mathbf{b} \tag{9.3}$$

Systems of linear equations with the nonzero right-hand side coefficients $b_i$ have a unique solution when the determinant of the matrix $\mathbf{A}$ is not equal to zero. If all the coefficients $b_i$ are equal to zero, then the solution exists if, and only if, the determinant of the matrix $\mathbf{A}$ is zero.

## 9.2 Gaussian elimination

For systems with a small number of equations, the analytic solutions can be easily found. In school you were already solving linear systems with two equations using the elimination technique. For example, expressing the first unknown $x_1$ from the first equation and substituting in the second equation gives a single equation with one unknown $x_2$. Because there is no such operator as `elimination` either in C++ or in FORTRAN we should translate this procedure to an appropriate numerical method for solving systems of linear equations. For clarity we will consider a system with three linear equations:

$$a_{11}x_1 + a_{12}x_2 + a_{13}x_3 = b_1$$
$$a_{21}x_1 + a_{22}x_2 + a_{23}x_3 = b_2 \tag{9.4}$$
$$a_{31}x_1 + a_{32}x_2 + a_{33}x_3 = b_3$$

Generalization for $n$ equations is then straightforward. Let us subtract the first equation, multiplied by the coefficient $a_{21}/a_{11}$, from the second equation, and multiplied by the coefficient $a_{31}/a_{11}$ from the third equation. The system (9.4) will transform into the following

$$a_{11}x_1 + a_{12}x_2 + a_{13}x_3 = b_1$$
$$0 + a'_{22}x_2 + a'_{23}x_3 = b'_2 \tag{9.5}$$
$$0 + a'_{32}x_2 + a'_{33}x_3 = b'_3$$

where the new coefficients $a'_{ij} = a_{ij} - a_{1j}a_{i1}/a_{11}$ and $b'_i = b_i - b_1 a_{i1}/a_{11}$, $i = 2, \ldots, n$, $j = 1, \ldots, n$. We may notice that the unknown $x_1$ has been eliminated from the last two equations. If we ignore the first equation, then the last two equations form a system of two equations with two unknowns. Repeating the same operation for the last two equations gives

$$a_{11}x_1 + a_{12}x_2 + a_{13}x_3 = b_1$$

$$0 + a'_{22}x_2 + a'_{23}x_3 = b'_2 \tag{9.6}$$

$$0 + 0 + a''_{33}x_3 = b''_3$$

with $a''_{ij} = a'_{ij} - a'_{2j}a'_{i2}/a'_{22}$ and $b''_i = b'_i - b'_2 a'_{i2}/a'_{22}$, $i = 3, \ldots, n$, $j = 2, \ldots, n$. For a system with $n$ equations we repeat the procedure of this forward elimination $n - 1$ times. From the last equation follows $x_3 = b''_3/a''_{33}$. Doing back substitution we will find $x_2$ and then $x_1$. This direct method to find solutions for a system of linear equations by successive eliminations is known as Gaussian elimination.

This method appears to be very simple and effective. However, in our consideration we missed several important points: zero diagonal elements, round-off errors, ill conditioned systems, and computational time. What if one of the diagonal elements is zero? The answer is clear: the procedure will fail. Nevertheless, the system may have a unique solution. The problem may be solved by interchanging the rows of the system, pushing elements which are zero off the diagonal. This is called the partial pivoting procedure. Moreover, reordering the system in a way when $a_{11} > a_{22} > a_{33} > \cdots > a_{nn}$ would increase the efficiency of this method. This is the issue of complete pivoting.

For each new elimination, the Gaussian method employs results from the previous ones. This procedure accumulates round-off errors and thus for large systems you may get a wrong numerical solution. It is highly recommended to check your solutions by direct substitution of $x_1, x_2, \ldots, x_n$ into the original system of linear equations. How can we reduce the round-off errors? Usually, complete pivoting may be very efficient. Besides scaling, multiplication of the $i$th equation by a constant $c_i$ may also help in improving accuracy.

Another possible disaster in the waiting is the case of ill conditioned systems. For such systems a small change in coefficients will produce large changes in the result. In particular, this situation occurs when the determinant for $\mathbf{A}$ is close to zero. Then the solution may be very unstable. An additional issue is time. As the number of equations in the system increases, the computation time grows nonlinearly. Systems with hundreds, even thousands, of equations are common in physics. And you may face a problem of waiting for weeks, if not years, to get an output from your computer. Sometimes iterative methods can help to increase speed, but generally they are less accurate.

## 9.3 Standard libraries

At this moment you may be overwhelmed by terminology such as: "pivoting," "scaling," "round-off errors," etc. How much experience and time do we need to understand the basic variations of just Gaussian elimination? How about other methods? Jordan elimination, LU decomposition and singular value decomposition are all widely used techniques. In addition, there are numerous methods for special forms of equations: symmetrical, tridiagonal, sparse systems. We believe that the most powerful method to solve most systems of linear equation is the use of "standard libraries." In Appendix B.1 and B.2 you will find a list of some of the most popular libraries on the Web with many robust programs for solving systems of linear equations. Specifically, the LAPACK library is a very large linear algebra package with hundreds of programs. However, you have to be careful in selecting a program that is right for your specific system of linear equations.

## 9.4 Eigenvalue problem

The eigenvalue problem is the key to structure calculations for quantum systems in atomic, molecular, nuclear and solid state physics. Mathematically, the eigenvalue problem may be written as

$$\mathbf{A} \cdot \mathbf{x} = \lambda \mathbf{x} \qquad (9.7)$$

or

$$
\begin{aligned}
a_{11}x_1 + a_{12}x_2 + \cdots + a_{1n}x_n &= \lambda x_1 \\
a_{21}x_1 + a_{22}x_2 + \cdots + a_{2n}x_n &= \lambda x_2 \\
&\cdots\cdots\cdots\cdots\cdots\cdots\cdots\cdots\cdots\cdots \\
a_{n1}x_1 + a_{n2}x_2 + \cdots + a_{nn}x_n &= \lambda x_n
\end{aligned}
\qquad (9.8)
$$

The equations above look like a linear system of equations (9.1). However, there is a substantial difference between the eigenvalue problem and the linear system of equations. For the eigenvalue problem the coefficients $\lambda$ are unknown and solutions for the system (9.8) exist only for specific values of $\lambda$. These values are called *eigenvalues*.

Regrouping terms in the system (9.8) leads to

$$(a_{11} - \lambda)x_1 + a_{12}x_2 + \cdots + a_{1n}x_n = 0$$

$$a_{21}x_1 + (a_{22} - \lambda)x_2 + \cdots + a_{2n}x_n = 0$$

$$\dots\dots\dots\dots\dots\dots\dots\dots\dots\dots\dots\dots$$

$$a_{n1}x_1 + a_{n2}x_2 + \cdots + (a_{nn} - \lambda)x_n = 0$$

(9.9)

Introducing a unit matrix $\mathbf{I}$, which is defined as

$$I = \begin{pmatrix} 1 & 0 & \cdots & 0 \\ 0 & 1 & \cdots & 0 \\ \cdots & \cdots & \cdots & \cdots \\ 0 & 0 & \cdots & 1 \end{pmatrix}$$

(9.10)

one can rewrite the system of linear equations (9.9) in the following form:

$$(\mathbf{A} - \lambda\mathbf{I}) \cdot \mathbf{x} = 0$$

(9.11)

Solutions for the system (9.11) exist if, and only if, the determinant of the matrix $\mathbf{A} - \lambda\mathbf{I}$ is zero

$$\det|\mathbf{A} - \lambda\mathbf{I}| = 0$$

(9.12)

For an $(n \times n)$ matrix the equation above would give a polynomial in $\lambda$ of degree $n$

$$c_n\lambda^n + c_{n-1}\lambda^{n-1} + \cdots + c_1\lambda + c_0 = 0$$

(9.13)

The coefficients $c$ are determined through the matrix elements $a_{ij}$ by the definition for the matrix determinant. This polynomial equation is known as the characteristic equation of the matrix $\mathbf{A}$. Roots of this equation would give the required eigenvalues. In physics, we often deal with either symmetric $a_{ij} = a_{ji}$ or Hermitian $a_{ij} = a_{ji}^*$ matrices ($a^*$ stands for the complex conjugate element). It is important to know that all the eigenvalues for these matrices are real.

In Chapter 7 we discussed methods for solving nonlinear equations. For matrices with small $n$ these methods may be applied to finding the eigenvalues $\lambda$ from equation (9.13). Once we have determined the eigenvalues, we may solve the system of linear equations (9.9) to find a set of solutions $\mathbf{x} = \{x_1, x_2, \ldots, x_n\}$ for each value of $\lambda$. These solutions are called the *eigenvectors*. For Hermitian matrices, the eigenvectors corresponding to distinct eigenvalues are orthogonal.

In general, the scheme above for solving the eigenvalue problem looks very straightforward. However, this scheme becomes unstable as the size of the matrix increases. The standard libraries have many robust and stable computer programs for solving the eigenvalue problem. In particular, programs based on the Faddeev–Leverrier method are very popular and successful in atomic and molecular structure calculations. The Lanczos algorithm is a good choice for large and sparse matrices which are common in many-body problems.

## 9.5 Exercises

1. Write a program that implements the Gaussian elimination method for solving a system of linear equations. Treat $n$ as an input parameter.
2. Apply the program to solve the set of equations

$$\begin{pmatrix} 2 & 3 & -2 \\ 1 & -6 & 4 \\ 4 & -1 & 6 \end{pmatrix} \begin{pmatrix} x_1 \\ x_2 \\ x_3 \end{pmatrix} = \begin{pmatrix} 2 \\ 4 \\ 6 \end{pmatrix}$$

3. Compare accuracy of the Gaussian elimination method with a program from a standard library for solutions of the following system of equations:

$$\begin{pmatrix} 1 & \frac{1}{2} & \frac{1}{3} & \frac{1}{4} \\ \frac{1}{2} & \frac{1}{3} & \frac{1}{4} & \frac{1}{5} \\ \frac{1}{3} & \frac{1}{4} & \frac{1}{5} & \frac{1}{6} \\ \frac{1}{4} & \frac{1}{5} & \frac{1}{6} & \frac{1}{7} \end{pmatrix} \begin{pmatrix} x_1 \\ x_2 \\ x_3 \\ x_4 \end{pmatrix} = \begin{pmatrix} 4 \\ 3 \\ 2 \\ 1 \end{pmatrix}$$

4. Write a program that calculates eigenvalues for an $n \times n$ matrix. Implement the program to find all eigenvalues of the matrix:

$$\begin{pmatrix} 1 & 2 & 3 \\ 2 & 2 & -2 \\ 3 & -2 & 1 \end{pmatrix}$$

Using a program from standard libraries find all eigenvalues and eigenvectors of the matrix above. Compare the results with your program for eigenvalues

# Chapter 10
# Random processes and Monte Carlo simulation

## 10.1 Random processes in science

In science as well as in everyday life we constantly encounter situations and processes which are stochastic in nature, i.e., we cannot predict from the observation of one event how the next event with the same initial starting conditions will come out. Say you have a coin which you throw in the air. The only prediction about the outcome you can make is to say that you have a 50% chance that the coin will land on its tail. After recording the first event, which was a head, you toss it up again. The chance that the coin will land on its tail is still 50%. As every statistics teacher has taught you, you need to execute these throws a large number of times, to get an accurate distribution. The same is true when you roll a die. The only prediction you can make is that in a large number of throws each number has a probability of 1/6. Now assume that you have 10 dice, and you are to record all distributions in 10 000 throws. Of course probability theory will predict the outcome for any combination you wish to single out; for example, the chances to get all "1s" in one throw (1/6). However, actually trying to do this experiment would keep you busy for quite some time (and in addition you would not learn anything new). This is where the so-called "Monte Carlo" simulation technique comes into play. The idea is that you let the computer throw the coin for you and also record the outcome, which you can then plot. In this scheme you simply need a program which generates randomly a variable, which can have only two values, say 1 and 0. Having the computer do this 10 000 times and record the outcome each time will give you the proper distribution. But, how do you write a program which creates a random number? And how do you know that such a number is truly random?

## 10.2 Random number generators

Before we discuss how to create random numbers with computers, let us make the following announcement:

There is no true computer-generated random number!

A computer has only a finite precision because of the limited number of bits it has to represent a number. Eventually the sequence of numbers from the generator will repeat itself. Say you record all the numbers produced by the computer and carefully look at the series. At some point you will notice that the numbers repeat themselves in exactly the same sequence. This is called the period of the generator. The following example illustrates why you have to be concerned about this. Assume you write a program, which simulates a simple process like rolling one die and recording the outcome. In order to have around 10 000 occurrences for each possible outcome, therefore giving you a statistical error of 1%, you would have to have the computer generate 60 000 events. If your random number generator had a period of only 10 000, you actually would have only the statistical accuracy corresponding to 10 000 and not 60 000, because you just repeat the same sequence six times. The tricky part, however, is that unless you check your generator for the length of the period, you will not be aware of this problem.

Other issues which you have to be concerned about are uniformness over the range of numbers and possible correlations between numbers in the sequence. The ideal random number generator should distribute its numbers evenly over an interval. Otherwise you will bias your simulation with the distribution you have originally generated.

The most widely known random number generators (RNG) are the so-called *linear congruential generators* and are based on a congruence relationship given by:

$$I_{i+1} = (aI_i + c) \bmod(m) \tag{10.1}$$

where $m$ is the modulus and $a$ and $c$ are constants chosen by the programmer. This recursive relation generates the next random integer from the previous one. Herein lies one of the problems of this sequence. If by chance the new number $I_{i+1}$ has already shown up before, the whole sequence repeats and you have reached the end of the period. Another problem we want to mention in

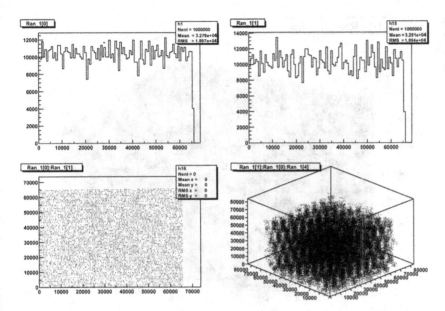

**Figure 10.1** The results from a unsatisfactory random number generator.

passing is the problem of correlations. If the constants in the generator are not carefully selected, you will end up with a situation where you could predict a later random number by looking at a previous one. Below is a small routine which demonstrates the problems of a simple random number generator:

```
int rand_ak(int &random)
{
// the generator I_(i+1)=aI_(i)+c (mod m)
int a=65;
int c=319;
int m=65537;

random=(a*random+c) % m;

return (random);

}
```

The main program creates eight different series of random numbers. These can be plotted independently (like the upper two panels in Figure 10.1) or displayed one variable against one from a different series (see lower left panel). In the lower right panel we show a three-dimensional plot of three variables. Already in the one dimensional case the problems are clearly recognizable. There are fluctuations in the distributions which are way outside

**Figure 10.2** The results
from the Linux random
number generator.

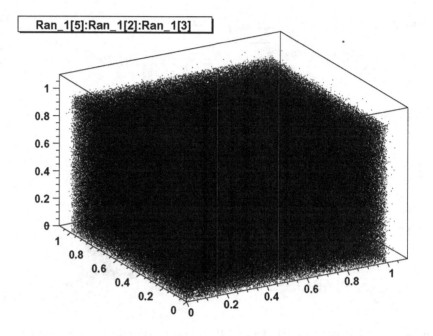

**Figure 10.2** The results from the Linux random number generator.

of any statistical variation, indicating that some numbers show up more often than others. Graphically, the most intriguing is the three dimensional case. There are clear lines visible, which reflect a strong correlation among the three random numbers.

An acceptable generator will fill a three dimensional cube completely homogeneous as is the case in the next figure 10.2:

This distribution was generated from the Linux generator rand3 (described in man pages **rand(3)**). The number generated is divided by **RAND_MAX** (the largest generated number; provided by the system), so that the interval is now from zero to one.

## 10.3 The random walk

To illuminate the power of simulations with the computer we start with a very simple problem, namely the random walk in one dimension and then gradually add more complexity to it. This is closely related to the phenomena of diffusion.

Suppose you live in a one dimensional world and you only can take steps either to the left or to the right, like on a very narrow mountain path. The step length in either direction is the same and you choose randomly if you

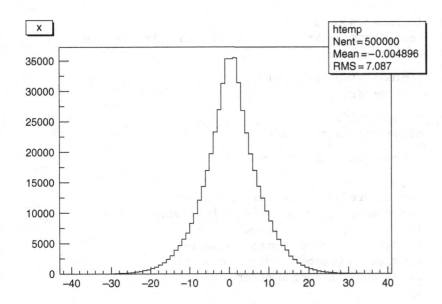

**Figure 10.3** The position distribution for 5000 walks and each 100 steps.

want to go left or right. This is also referred to as the drunkard's walk. The question then is, where will you be after $N$ steps?

$$x = ml \qquad (10.2)$$

where $l$ is the length of the step and $m$ is such that $-N \le m \le N$. Because the probability for stepping to the right is the same as stepping to the left, the most likely position for several walks will be at $x = 0$ and the distribution for the positions will be binomial.

The following program has 5000 different walks in it and each walk has 100 steps (Figure 10.3). This program is also a nice illustration for the power of ROOT trees, which will allow you to analyze the different walks.

```
// rnd_walk1.cxx
// Random walk in one dimension
// ak 4/30/2002
#include <iostream>
#include <fstream>
#include <math.h>
#include <iomanip>
#include <stdio.h>
#include <stdlib.h>
#include <fcntl.h>
```

```
#include "TROOT.h"
#include "TTree.h"          // we will use a tree to store
our values

#include "TApplication.h"
#include "TFile.h"

int rand(void);
void srand(unsigned int);

void main(int argc, char **argv)
{
 //structure for our random walk variables
 struct walk_t { int x; // the position after a step
  int nstep; // the step number
  int left; // number of steps to the left
  int right; // number of steps to the right
  int jloop; // number of outer loops
  };
 walk_t walk;

 int i_loop;              //inner loop
 int j_loop;              //outer loop
 int jloop_max=5000; // the max number of different trials

   unsigned int seed = 68910 ; // here is the starting
   value or seed

     int loop_max=100; // the maximum number of steps

   double rnd;

 TROOT root("hello", "computational physics");

 // initialize root
 TApplication theApp("App", &argc, argv);

 //open output file
 TFile *out_file = new TFile("rnd_walk68910.root",
 "RECREATE", "example of random random walk"); //
 create root file

 // Declare tree
 TTree*ran_walk = new TTree("ran_walk", "tree with
 random \index{random} walk variables");

   ran_walk->Branch("walk",&walk.x,
 "x/I:nstep/I:left/I:right/I:jloop/I");

 // set seed value
```

```
srand(seed);
// the outer loop, trying the walk jloop times
 for(j_loop=0;j_loop < jloop_max ;j_loop= j_loop+1)
 {
walk.x=0;
walk.nstep=0;
walk.left=0;
walk.right=0;
walk.jloop=j_loop+1;
 for(i_loop=0;i_loop < loop_max ;i_loop= i_loop+1)
 {
// here we get the step
rnd=double(rand())/double(RAND_MAX);
if((rnd-.5)<=0.)
{
 walk.x=walk.x-1;
 walk.left=walk.left+1;
 }
  else
 {
   walk.x=walk.x+1;
 walk.right=walk.right+1;
 }
 walk.nstep=walk.nstep+1;
 // fill the tree
  ran_walk->Fill();

 }
  }
  out_file->Write();

 }
```

To analyze the ROOT file `rnd_walk68910.root` you have created, you start up ROOT and read in the file:

```
Welcome to the ROOT
root [0] TFile*f = new TFile ("rnd_walk68910.root")
root [1] TTree*mytree = ran_walk
root [2] mytree->ls()
OBJ: TTree ran_walk tree with random\indexrandom walk
variables : 0
*************************************************************
```

```
*Tree        :ran_walk   : tree with random\indexrandom walk
                              variables
*Entries :   500000 : Total =     10018021 bytes File Size =
                          2852491                              *
*         :            : Tree compression factor = 3.52       *
***********************************************************
* Br     0 :walk       : x/I:nstep/I:left/I:right/I:jloop/I  *
* Entries :   500000 : Total   Size=   10011931 bytes        *
                          File  Size =   2846401              *
* Baskets :       313 : Basket Size=        32000 bytes      *
                          Compression=   3.52                 *
*.............................................................*
root [4]
```

With the `mytree->Print()` statement you get a list of the tree and the branches, in our case only one branch, called "walk" with several leaves. The next command will draw the positions for all the different walks, representing a binomial distribution:

$$root[4]mytree-> Print()$$

Now in order to look at the different walks, say for instance walk number 5, we set the condition `jloop==5`, on the analysis:

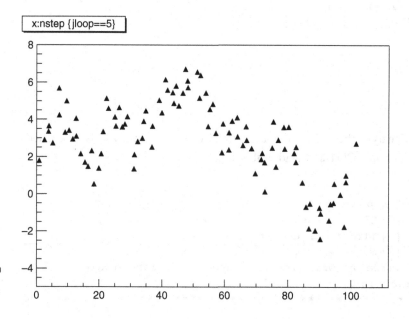

**Figure 10.4** The $x$ distribution as a function of the number of steps for trial number 5.

```
root [5] mytree->Draw("x:nstep", "jloop==5", "P")
```

This gives the random walk for the fifth trial (Figure 10.4).

In Appendix D.3 you will find a program which describes the random walk in two dimensions.

## 10.4 Random numbers for nonuniform distributions

A lot of situations in physics and science require random numbers which are not distributed uniformly. For example, a typical beam from a particle accelerator has an energy distribution around its mean energy, which in some cases can be approximated by a Gaussian distribution. The challenge then is to generate random numbers which follow this probability. In addition, you now have to deal with intervals which are different from $[0, 1)$. There are several different options to produce a specific distribution with a uniform random number generator.

### Acceptance and rejection method

This is probably the easiest method to implement, however it might not be the fastest. This procedure is related to the Monte Carlo integration technique. Suppose you want to generate numbers which follow a certain distribution $f(x)$, with $x \in [a, b]$. The most straightforward way to achieve this is to throw all events away which are outside of $f(x)$ in a two dimensional plane. If, for example, your distribution is $f(x) = \sin(x)$ in $[0, 180)$, you would use as your $x$ variable a uniformly distributed number between 0 and 180. As your $y$-coordinate for your plane you would use a second independent variable between 0 and 1. If your random generator is uniform, then initially this plane is filled uniformly in the two dimensions. If you now reject all the $(x, y)$ points where $y > f(x)$ (in our case $\sin(x)$), the remaining accepted $x$ will be distributed according to $\sin(x)$ (see Appendix D.4).

As an example of this we will discuss the Drell–Yan process in high energy physics [9]. A nucleon, proton or neutron, is made up of what is called three valence quarks, and a "sea" of quark–anti-quark pairs, which you can consider as virtual excitations, the same way as a photon can be described by an electron–positron pair. As you might know, there exist six different quarks and the corresponding anti-quarks, but for the case of the nucleon we will only deal with the $u$ and $d$ quarks. Quarks have fractional charge, the $u$

quark has $Q_u = 2/3$ and the $d$ has $Q_d = -1/3$. It is immediately clear that the proton consists of two $u$ and one $d$ quarks because the total charge is $Q_p = 1$. Similarly, the neutron has two $d$ and one $u$ quarks. If you combine a quark and an anti-quark together, they will annihilate into a virtual photon, which then can decay either back into two quarks or into two leptons, like an electron–positron or muon–anti-muon pair.

The Drell–Yan process is exactly this annihilation of a quark–anti-quark pair into a heavy virtual photon, which then decays into a $\mu^-\mu^+$ pair.

In order to study this reaction you collide two proton beams of high energy with each other and during the interaction one of the quarks from one proton will annihilate with an anti-quark from the "sea" of the other. The cross-section for this process can be written down in the following way:

$$\frac{\mathrm{d}^2\sigma}{x_1 x_2} = \text{constant} * \frac{1}{M^2} \sum_i e_i^2 \left[ f_i^A(x_1)\bar{f}_i^B(x_2) + \bar{f}_i^A(x_1)f_i^B(x_2) \right] \tag{10.3}$$

Even though this formula looks somewhat intimidating it is actually rather straightforward. The first terms to be defined are the $x_1$ and $x_2$. The subscripts 1, 2 refer to the two quarks from the beams. The $x$ is the Bjorken $x$, which expresses the fraction of the momentum of the nucleon carried by the respective quarks. Because a nucleon has three valence quarks, each of them will carry on average a third of the nucleon momentum. However, as for a nucleon bound in a nucleus, the quark has a momentum distribution, which can be between 0 (at rest) and 1 (carrying all the momentum). In Figure 10.5 we show the parameterization found from experiments [10] for the three different quark types:

$$u_v(x) = 2.13\sqrt{x}(1-x)^{2.8} \tag{10.4}$$

$$d_v(x) = 1.26\sqrt{x}(1-x)^{3.8} \tag{10.5}$$

$$\overline{g_{\text{sea}}} = 0.27(1-x)^{8.1} \tag{10.6}$$

In Equation (10.3) the $f_i$ corresponds to the two valence distributions, while $\bar{f}_i$ represents the sea quark distribution. The term $e_i^2$ stems from the fractional charge of the quarks and the index $i$ goes from 1 to 3 for the three valence quarks which can be involved. The last variable to be explained is $M^2$. This is the square of the invariant mass of the produced di-lepton pair and is a function of the total energy $S$ and the momentum fractions of the participating quarks $x_1$ and $x_2$:

$$M^2 = Sx_1 x_2 \tag{10.7}$$

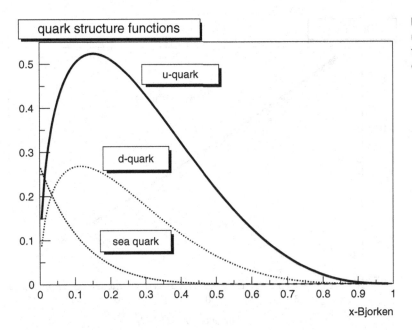

**Figure 10.5** The quark momentum distributions for the *u*, *d* and the sea of quarks.

In the program `drell_yan1.cxx`, the Drell–Yan cross-section is calculated using the rejection method and a beam energy of 25 GeV. The result is shown in Figure 10.6.

## Inversion method

Suppose you want to simulate some decay process, like the decay of an ensemble of radioactive nuclei. In the case of a nuclear decay, we could for instance choose the decay time at random from the given decay law

$$f(t) = \exp(-\lambda t) \tag{10.8}$$

This means we would like to choose a variable which follows an exponential distribution. In this case there is a much more efficient and faster method available, namely the inverse transform method. We will briefly describe this in the following section.

The usual situation is that we have a uniform distribution $x$ available in the interval [0, 1). The probability of finding a number in the interval d$x$ around $x$ is then given by

$$p(x)\, \mathrm{d}x = \mathrm{d}x \quad \text{for } 0 \leq x < 1$$
$$= 0 \quad \text{otherwise} \tag{10.9}$$

**Figure 10.6** The Drell–Yan
cross-section as a
function of the
invariant mass.

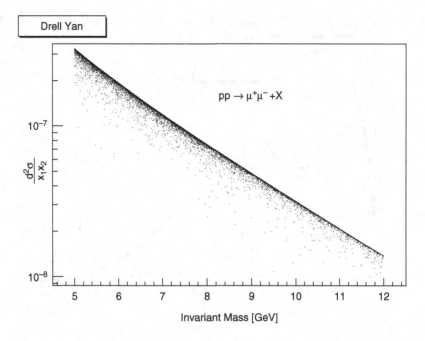

The probability has to be normalized, i.e.,

$$\int_{\infty}^{-\infty} p(x)\,\mathrm{d}x = 1 \tag{10.10}$$

Our goal is to obtain a distribution $P(y)\,\mathrm{d}y = \mathrm{e}^{-y}\,\mathrm{d}y$, which implies that we have to find the proper transformation. If we require that $P(y)\,\mathrm{d}y$ has the same normalization as $p(x)$ we can write

$$P(y)\,\mathrm{d}y = p(x)\,\mathrm{d}x \tag{10.11}$$

This together with Equations (10.9) and (10.10) allows us to write

$$\mathrm{e}^{-y} = x \tag{10.12}$$

which solving for $y$ leads to:

$$y(x) = -\ln(x) \tag{10.13}$$

This is a very fast and efficient way to produce an exponentially decaying distribution. This method, however, will work only if the function we are trying to use for a distribution has an inverse:

$$x = F(y) = \int P(y)\,\mathrm{d}y$$
$$y = F^{-1}(x)$$

## Additional distributions

Other distributions, which are important in physics, are the normal or Gaussian, the Lorentz, and the Landau distributions. In *Handbook of Mathematical Functions* [6] you will find these and other examples. In addition, ROOT provides most of the common distributions and will give you any random distribution according to a user defined function.

```
// Define function to be sampled
TF1 *f1=new TF1("f1", "exp(-x)",0.,1.);

//x_random is now distributed according to exp(-x)
x_random=f1.GetRandom();
```

## 10.5 Monte Carlo integration

This section deals more with the mathematical application of random numbers than with a physical one. We describe how Monte Carlo methods can be applied to the integration of complicated functions. These procedures are especially useful when you are trying to integrate multi-dimensional integrals.

The best way to understand the principle of Monte Carlo integration is by using a geometrical description. Imagine you are charged with the task of determining the area of the polygon in Figure 10.7. One way would be to create little squares and count how many squares can be put into the shape. However, suppose you have a shotgun, which with every shot you take, sprays the whole area of the rectangle homogeneously and, in addition, you know how many pellets in total came out from one round. In order to determine the area, you count how many pellets went into the polygon area and compare them with the total number of pellets in the complete area. Clearly a rectangular spray from a shotgun is not very likely, but a circular area is probably a good approximation. The important part is that the enclosing area is easy to calculate and the inside is completely contained

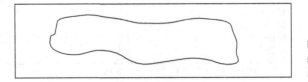

**Figure 10.7** A complicated shape bounded by a rectangular box.

in this outer shape. The most commonly used example demonstrating this is to determine the value of $\pi$. The area of a circle is $\pi r^2$, which for a circle of unit radius is equal to $\pi$. To compute this with the Monte Carlo integration method, you limit yourself to one quarter of the unit circle, and throw two random variables for $x$ and $y$. Any time $y$ is less than $\sqrt{(1-x^2)}$ you are inside the circle and count this as a hit. The following code fragment illustrates this:

```
for(i_loop=0;i_loop < i_loop_max ;i_loop= i_loop+1)
{
    x=double(rand())/double(RAND_MAX);
    y=double(rand())/double(RAND_MAX);

    if(y<=sqrt(1-pow(x,2)))
    {
      hit=hit+1; // the point is within the circle
      }
}
pi_result=4*hit/i_loop_max; // We have only used one quarter
```

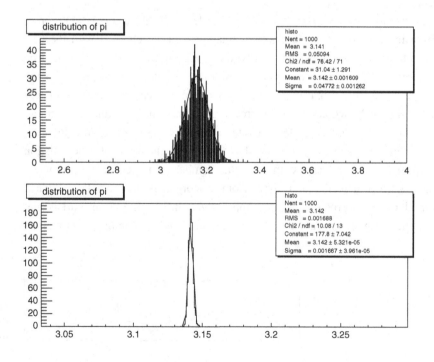

**Figure 10.8** The distribution of the value of $\pi$ for 1000 throws (upper) and 1 000 000 throws (lower panel).

Before we continue, we need to discuss how to estimate the accuracy and error of the Monte Carlo integration. In Figure 10.8 we have plotted the calculation of $\pi$ for two different cases. Both panels evaluated $\pi$. However, in the upper panel each integral was determined from 1000 random points, while in the lower panel we used 1 000 000 points. (Note the different scales for the axes.) The distribution has a Gaussian shape, which is often the case for random processes.

If you take a closer look at the $\sigma$ for both Gaussian fits, you will notice that $\sigma_{lower} \sim \sigma_{upper}/\sqrt{(1000)}$. The standard deviation can be written as

$$\sigma = \sqrt{\frac{\frac{1}{N}\Sigma f^2(x_i) - (\frac{1}{N}\Sigma f(x_i))^2}{N}} \qquad (10.14)$$

which immediately shows how $\sigma \propto N^{-1/2}$. In order to reduce the dispersion of our integral by 2 we need to throw four times as many points.

## 10.6 Exercises

1. Pi-mesons are unstable particles with a rest mass of $m_\pi \approx 140$ MeV/$c^2$. Their lifetime in the rest system is $\tau = 2.6^{-8}$ s. If their kinetic energy is 200 MeV, write a program which will simulate how many pions will survive after having traveled 20 m. Start with an initial sample of $10^8$ pions. Assume that the pions are monoenergetic.
2. Modify your program in such a way that the pions are not monoenergetic, but have a Gaussian energy distribution around 200 MeV with $\sigma = 50$ MeV.
3. Use Monte Carlo integration to calculate

$$\int_0^1 \frac{\ln(x)}{1-x}dx$$

# References

[1] Rene Brun and Fons Rademakers, ROOT – an object oriented data analysis framework, *Proceedings AIHENP'96 Workshop, Lausanne, September 1996, Nucl. Instrum. Methods Phys. Res.* **A 389** (1997) 81–86.

[2] Brian W. Kernighan and Dennis M. Ritchie, *The C Programming Language*. Englewood Cliffs, NJ: Prentice-Hall, 1978.

[3] Bjarne Stroustrup, *The C++ Programming Language*. New York: Addison-Wesley, 1986.

[4] S. Garfinkel, D. Weise and S. Strassmann, editors, *The UNIX-Haters Handbook*, San Mateo, CA: IDG Books, 1994.

[5] W. H. Press, S. A. Teukolsky, W. T. Vetterlin and B. P. Flannery, *Numerical Recipes in C*, p. 25. Cambridge: Cambridge University Press, 2nd edition, 1995.

[6] M. Abramowitz and I. A. Stegun, *Handbook of Mathematical Functions*, New York: Dover Publications, 1964.

[7] E. Fehlberg, Low order classical Runge–Kutta formulas with step-size control and their application to some heat transfer problems, NASA TR R-315, 1969.

[8] J. R. Cash and A. H. Karp, *ACM Trans. Math. Software* **16** (1990) 201.

[9] S. Drell and T. M. Yan, *Phys. Rev. Lett.* **25** (1970) 316.

[10] D. Antreasyan *et al.*, *Phys. Rev. Lett.* **48** (1982) 302.

[11] *ROOT Users' Guide*, CERN, 2001.

# Appendix A
# The ROOT system

## A.1 What is ROOT

Root is a very powerful data analysis package which was developed and written at CERN, the European high energy accelerator lab. It consists of a library of C++ classes and functions for histogramming, data manipulation (like fitting and data reduction) and storage tools. There are two ways to use ROOT. One way is by calling the appropriate functions from your own program. The second way relies on a C++ interpreter and a graphical user interface, allowing you to display and manipulate data interactively. Originally written for high energy physics, it is today also widely used in other fields, including astrophysics, neural network applications, and financial institutions on Wall Street.

## A.2 The ROOT basics

ROOT in the word of its authors is an "object oriented framework" [11]. Instead of having to write your own graphics routines, or modify existing programs to suit your needs, a framework provides you with functions and routines which are already well tested. ROOT, having been developed by physicists for physics problems, gives you a large class of routines which you can just use "out of the box."

One of the main advantages of ROOT is that it is running and supported on all the major UNIX platforms, as well as on Windows XP and MacOS. This widespread availability makes it the perfect choice, reducing the chances of having to learn a new package every time you change computers. This appendix is a short introduction. In order to take advantage of the full potential of the ROOT package, we refer you to the ROOT users manual, which you can find on the ROOT Web page http://root.cern.ch/root/RootDoc.html. Also check out the main page which has lots of useful links. The most important of these is the **Reference Guide**, which has a listing of all the classes and documentation on how to use them.

The only way of learning a new language or program is to use it. This appendix heavily relies on small exercises and problems, which you solve either in the lab or at home with your computer. The first thing you have to do is find out whether the ROOT environment is defined or whether you have to define it (assuming that it has been installed; to obtain the distribution, look in Appendix B.3). By typing **echo $ROOTSYS**, the system will tell you whether this environment variable has been defined. Once you have convinced yourself that ROOT indeed is defined, we can explore the system. In this course, we will use ROOT with the graphical interface, reducing to a minimum the need for calls to the ROOT libraries. In most of the programs, the only reference to ROOT will be writing to a ROOT file and saving it. Once your program has finished (hopefully successfully), you will then start up ROOT.

## A.3 The first steps

This section introduces you to writing your output to a ROOT file, using a simple program, which will calculate the sin of an angle.

```
// program to calculate the sin for a given interval and step size
// this is the first step to using ROOT.
# include "iostream.h"
# include "math.h"

void main()
{
double xmin, xmax; // the lower and upper bounds of the interval
double xval; //the value where we calculate the sin
double step; // the step size for the angle increment
double result;
double const deg_rad = 3.141592653589793/180.; // converts deg to
radians

cout << "This programs calculates the sin for a range of angles \n
You have to give the lower and upper boundaries for the angles \n
and the step size \n";

cout << "give lower and upper angle limit";
  cin >> xmin >> xmax;
```

```
cout << "Give step size";
  cin >> step;

// here we loop over the angles

step = step*deg_rad; // convert the step size into radians
  xmin = xmin*deg_rad;
    xmax = xmax*deg_rad;

for (xval = xmin; xval<=xmax; xval+=step)
{
  result = sin(xval);
  cout << "The sin of" << xval/deg_rad <<"degrees is" <<result <<"\n";
}
}
```

This is a very simple program and will output the values for $\sin(x)$ onto your screen. This is not very useful if you want to see a whole list of values. The first thing you could do would be to redirect the output from the program into a file; i.e., `sin1 ¿ sin.dat` and then plot this output file with plotting software.

In the next step we will take this program and modify it in such a way that it will send the output to a ROOT file. In ROOT, because it deals with objects, these are also the quantities you write to a file. You can save any object you created in your program or your interactive section to a file, and later open this file again with either a program or the interactive version. However, before you can use these features you have to initialize the ROOT system in your program. You will also need to include the header files belonging to the specific class you want to use. There are two different ways to get the output into a file. One way is to use a histogram from the one dimensional histogram class **TH1** and using the **TGraph** class.

### Creating a histogram file

In this example we have used the TH1 class to create a histogram.

```
// program to calculate the sin for a given interval and step size
// this is the modified sin1 program now using the ROOT
system.

# include "iostream.h"
# include "math.h"
```

```
// Here starts the block of include files for ROOT
//*************************************************************//
#include "TROOT.h" // the main include file for the root system
#include "TFile.h" // include file for File I/O in root
#include "TH1.h" // To create histograms
//*************************************************************//

void main()
{
double xmin, xmax; // the lower and upper bounds of the interval
double xval; //the value where we calculate the sin
double step; // the step size for the angle increment
double result;
double const deg_rad = 3.141592653589793/180.; // converts deg to
radians
int nbin; // number of bins for histogram


//******************* ROOT*********************************
TROOT root ("hello", "the sine problem"); //initialize root
TFile *out_file = new TFile("sin_histo.root", "RECREATE",
"sin_histo");
// The RECREATE option, will create or overwrite the file if it already
exists.//
//***************ROOT*********************************

cout << "This program calculates the sin for a range of angles \n"
You have to give the lower and upper boundaries for the angles \n
and the step size \n";

cout << "give lower and upper angle limit";
  cin >> xmin >> xmax;
cout << "Give step size";
  cin >> step;

// In order to define a histogram, we need to know how many bins we will
   have.
// we calculate the number automatically from the upper and lower
   limits and divide by
// the step size.

nbin = abs (static_cast<int>((xmax-xmin)/step)) +1;

// Now define a pointer to a new histogram, which is defined for double

TH1D *hist1 = new TH1D("hist1", "sin(x)", nbin, xmin, xmax);

// here we loop over the angles
```

```
step = step*deg_rad; // convert the step size into radians
  xmin=xmin*deg_rad;
   xmax=xmax*deg_rad;
for(xval=xmin; xval<=xmax; xval+=step)
{
result=sin(xval);

//Here we fill the bins of the histogram;
  hist1->Fill(xval/deg_rad,result);
}

//And last we need to write the histogram to disk and close the file

  hist1->Write();
  out_file->Close();
}
```

As you can see, by adding a few statements, dealing with the number of bins and the ROOT classes, we are now ready to use this program to create graphical output. However, we need to change the simple compile command to include the ROOT libraries. Instead of a simple:

```
g++ -o sin1 sin1.cxx
```

as in the previous example, we have to tell the compiler what needs to be included, and where the necessary files can be found. A simple script to do this, called make_sin2 is shown in the next paragraph (this script uses the make facility):

```
# simple scriptfile to compile sin2
# First define the ROOT stuff
ROOTCFLAGS  = $(shell root-config --cflags)
ROOTLIBS    = $(shell root-config --libs)
ROOTGLIBS   = $(shell root-config --glibs)
CXXFLAGS    += $(ROOTCFLAGS)
LIBS        = $(ROOTLIBS)
GLIBS       = $(ROOTGLIBS)

sin2:
g$++$ -o sin2 sin2.cxx $(ROOTCFLAGS) $(ROOTGLIBS) $(ROOTGLIBS)
```

**Note**: g++ is indented by a **TAB**, which is required by the make facility, and belongs to the target **sin2**. To execute this command you would type *make -f make_sin2 sin2*

This will then produce an executable called sin2.

## Creating a file with a graph

In this example we are taking advantage of the **TGraph** class in root. In a graph you create pairs of *x* and *y* values and then plot them. You first decide how many points you want to include in your graph, dimension *x* and *y* accordingly, and simply create a new object. Be careful that you are not trying to create more points than you have dimensioned.

```
// program to calculate the sin for a given interval and step size
// this is the modified sin1 program now using the ROOT system.
// Contrary to sin2 this one uses graphs instead of histograms.

# include "iostream.h"
# include "math.h"

// Here starts the block of include files for ROOT
//*************************************************************//
#include "TROOT.h" // the main include file for the root system
#include "TFile.h" // include file for File I/O in root
#include "TGraph.h" // To create a graphics
//*************************************************************//

void main()
{
int array=100;
// In order for the graph to be used we need to have an array of x and y.
// We dimension it for 100, but then have to make sure later on that we
// are not running out of boundary of the array

double xmin, xmax; // the lower and upper bounds of the interval
double xval1 [array]; //the value where we calculate the sin
double step; // the step size for the angle increment
double xval;
double result[array];

double const deg_rad = 3.141592653589793/180.; // converts deg to
radians
int nbin; // number of bins for histogram
```

```
//****************** ROOT*********************************
TROOT root("hello", "the sine problem"); //initialize root
TFile * out_file = new TFile ("sin_graph.root", "RECREATE",
"sin_graph");
// The RECREATE option, will create or overwrite the file if it already
exists.//
//********************** ROOT**********************************
```

```
cout << " This programs calculates the sin for a range of angles \n"
You have to give the lower and upper boundaries for the angles
and the step size \n";
```

```
cout << " give lower and upper angle limit ";
   cin >> xmin >> xmax;
```

```
cout << " Give step size ";
   cin >> step;
```

```
// Here we will check that we do not have more points than we defined in
  the array.
```

```
nbin = abs(static_cast<int>((xmax-xmin)/step)) +1;
if(nbin>array)
{
    cout << " array size too small \n";
    out_file->Close();
    exit;
    }
```

```
// Now define a pointer to a new histogram, which is defined for double
```

```
// here we loop over the angles
```

```
step = step*deg_rad; // convert the step size into radians
   xmin=xmin* deg_rad;
    xmax=xmax* deg_rad;
     array=0;
```

```
for( xval=xmin ; xval<=xmax ;xval+=step)
{
  result[array] = sin(xval);
    xvall[array]=xval/deg_rad;
    ++array;
}
```

```
   // Here we create the graph
   TGraph * graph =new TGraph(array,xvall,result);
```

```
//And last we need to write the graph to disk and close the file
graph->Write();
out_file->Close();
}
```

Again you have to modify your compile script.

## A.4 Lab ROOT

In this lab section we will try to show you how easy it is to calculate and plot functions with ROOT. Now that you have created two files, sin_histo.root and sin_graph.root, we can explore ROOT. The first thing you need to do is get ROOT up: *root*. This will bring you to the command line environment.

```
  *******************************************

  *                                         *

  *      W E L C O M E  to  R O O T         *

  *                                         *

  *  Version  3.01/00     9 May 2001        *

  * You are welcome to visit our Web site   *

  *        http://root.cern.ch              *

  *                                         *

  *******************************************

Compiled with thread support.
CINT/ROOT C/C++ Interpreter version 5.14.83, Apr 5 2001
Type ? for help. Commands must be C++ statements.
Enclose multiple statements between { }.
root [0]
```

The first thing you want to create is a so-called canvas, which will be the place where you will see all your plots. Because ROOT brings you into a C-interpreter, you execute commands as you would in your C++ program.

The command for creating the canvas is

*TCanvas *tc = new TCanvas("tc," "my first canvas")*

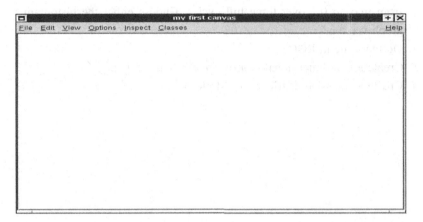

which will produce a plotting area in the upper left corner. To get control over your files, you should also start a browser, which will display the content of your directories with icons.

*TBrowser *tb = new TBrowser("tb," "my browser",500,500,500,400)*

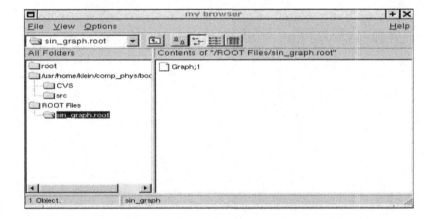

where we have given it $x$ and $y$ coordinates on screen and width and height of the browser. Now we are ready to draw our functions. In the browser, double click on one of the .root files, go to the ROOT files directory and double click on it again. It will then plot the chosen function.

## A.5 Exercises

1. Modify the sin program in such a way that the amplitude of the sin decays from max to 0.1 over three full cycles. Choose either the histogram or graph version.
2. Do the same in ROOT.
3. Create a two dimensional plot of $\sin(x)$ versus $\cos(x)$.
4. Create a Gaussian distribution and plot it.

# Appendix B
# Free scientific libraries

In the past, the dominant computer language for scientific computing was FORTRAN. This created a wealth of libraries with well tested routines, which address almost all numerical problems covered in this book. However, using these libraries "out of the box," without understanding their limitations and algorithms, is bound to get you into trouble one day. Because we are using C++, you will want to know how to call these FORTRAN routines from a C++ program. This will allow you to use these routines without having to rewrite them in C++.

## B.1 LAPACK

LAPACK: Linear Algebra PACKage.

> LAPACK is written in Fortran77 and provides routines for solving systems of simultaneous linear equations, least-squares solutions of linear systems of equations, eigenvalue problems, and singular value problems. The associated matrix factorizations (LU, Cholesky, QR, SVD, Schur, generalized Schur) are also provided, as are related computations such as reordering of the Schur factorizations and estimating condition numbers. Dense and banded matrices are handled, but not general sparse matrices. In all areas, similar functionality is provided for real and complex matrices, in both single and double precision.

From the Website http://www.netlib.org/lapack/index.html, where the whole package can be downloaded. There is now also a C++ version available, called LAPACK++.

## B.2 SLATEC

SLATEC is another very useful package which contains general pur-
pose mathematical and statistical programs. You will find this at
http://www.netlib.org/slatec/index.html.

## B.3 Where to obtain ROOT

An important aspect about ROOT to remember is the developer's philosophy:
"*Release early and release often.*" This requires you to check periodically
on the main website (http://root.cern.ch) for new releases. You can either
download the package in compiled form, or you can get the source code and
build the system yourself.

# Appendix C
# FORTRAN and C++

Before we can discuss how to call FORTRAN functions from C++, we need first to look at some of the differences.

1. The most important difference is that FORTRAN passes variables by reference, while C++, like C, passes by value. This in turn means that any FORTRAN routine expects to get a reference, so you have to make sure your program passes the variables appropriately.
2. When you use arrays in your program, you have to take into account that the first array element in FORTRAN is a(1), while in C++ it would be a[0]. Multi-dimensional arrays are stored differently in memory. Array A(3,5) in FORTRAN would be A(1,1), A(2,1), A(3,1), A(1,2), while in C++ the corresponding arrangement would be A(0,0), A(0,1), A(0,2), A(0,3), A(0,4), A(1,0) and so on.
3. C++ is a strongly typed language; if you do not define a variable your program will not compile. FORTRAN has by default an implicit type convention: any variable which starts with letters i through n is an integer, while any other variable is a real variable. Unless the programmer has used the "implicit none" statement any FORTRAN compiler will adhere to the standards.
4. FORTRAN is case insensitive, while C++ clearly distinguishes between Dummy and dummy.
5. Some compilers add an underscore to the name when they compile the program. You can list the contents of your library with **ar -t somelib.a** to see whether this is the case. Your compiler will usually have a switch (at least in UNIX) which is **-nosecondunderscore**.
6. How to call FORTRAN from C++ is strongly dependent on the C++ compiler you use. So if the Linux/g++ recipe does not work for you, you have to experiment.

## C.1 Calling FORTRAN from C++

The following shows an example of calling a routine from the SLATEC package, namely a Gaussian integration routine using Legendre polynomials. Only the statements which have to deal with this call are given:

```
// code snippet for function to integrate with gauss quadrature
// uses dgaus8.f from lapack.
// ak 9/2000
//
#include <iostream.h>
#include <fstream.h>
#include <math.h>
#include <iomanip.h>
#include <stdio.h>

extern "C" double dgaus8_(double (*func)(double*), double*,
double*, double* ,double*, int*); // This program uses the
slatec double precision gaussian
    //integration. You pass it a pointer to the function you
    // want to integrate.

main()
{
double integral=0.; // the calculated integral
  double x_low; //lower
  double x_high; //upper limits
int n_point; // number of integration points
double const deg2rad=3.14159265/180.; // converts degrees into
radians
double err=1.e-15; // tolerated error
int ierr=0;
.

.

.
dgaus8_(&func,&x_low,&x_high,&err,&integral,&ierr);

// compile and link
// g++ ~o gaus_int gaus_int.cxx -lslatec -llapack~-lg2c
```

Note the underscore at dgaus8_ and how all the variables are passed as pointers. At the end are included the compile and link commands for our Linux system. It links against SLATEC and LAPACK as well as libg2c.a. SLATEC calls some routines in LAPACK and the g2c library, which is the FORTRAN run time library.

# Appendix D
# Program listings

## D.1 Simple Euler

### Spring.cxx

```
// program to calculate the mass on a spring problem.
// m=k=1; therefore a=-x
// with the simple euler method
// ak 9/7/00
// all in double precision
#include <iostream.h>
#include <fstream.h>
#include <math.h>
#include <iomanip.h>
#include <stdio.h>
#include "TROOT.h"
#include "TApplication.h"
#include "TCanvas.h"
#include "TLine.h"
#include "TGraph.h"
 TROOT root ("hello", "Spring Problem"); // initialize root
int main (int argc, char **argv)
{
        double v_new[100]; // the forward value calculated
        double v_old=5.; // the previous value; for the first
        calculation
                        // the boundary condition
        double x_new[100]; // the forward x-value
        double x_old=0.; // the x value from previous
        double energy[100]; // the energy from mv**2/2+ kx**2/2
        double ti[100]; // array to hold the time
        double step =0.1; // the step size for the euler method
        double time=0. ; // the time

        int loop = 0; // the loop counter

        // first the root stuff
        TApplication theApp("App", &argc, argv);
```

```
TCanvas *c = new TCanvas ("c", "Hyperbola", 600, 600);
          //* create Canvas

// create the graphs //

while (loop <100)

     {
     V_new[loop]=v_old-step*x_old; // calculate the
     forward velocity
     x_new[loop]=x_old+step*v_old; // calculate the
     forward position
     energy[loop]=pow(x_new[loop],2)/2+pow(v_new
     [loop],2)/2.; // energy conservation
     energy[loop]=energy[loop]/5. ;// rescaled for
     plotting purposes

     time=time+step; // move forward in time by
     the step
     v_old=v_new[loop]; // the new value becomes now
     the old
     x_old=x_new[loop];
     ti[loop]=time;
     time=time+step; // move forward in time by
     the step
     loop=loop+1;
     }

TGraph *root1 = new TGraph(100,ti,x_new);
TGraph *root2 = new TGraph(100,ti,v_new);
TGraph *root3 = new TGraph(100,ti,energy);
     root1->SetTitle("Spring with simple Euler
     Method");

     root1->SetLineColor(1);
     root1->Draw("AC");
     root2->SetLineWidth(2);
     root2->SetLineColor(2);
     root2->Draw("C");
     root2->SetLineWidth(2);
     root3->SetLineColor(4);
     root3->Draw("C");

   theApp.Run();

}

// compile g++ -o spring spring.cxx with root libraries
```

## D.2 Runge–Kutta program

rk4.cxx

```cpp
// function to calculate step through ODE by
// 4th and 5th order Runge Kutta Fehlberg.
// calculates both orders and gives difference back
// ak 2/2/2001
// all in double precision
#include <iostream.h>
#include <fstream.h>
#include <math.h>
#include <iomanip.h>
#include <stdio.h>
void deri( double x, double y[], double f[], double omega2,
double gam);
double omega2;
double gam;
int len=2;
        double y[2],y_old[2];
        double f0[2], f1[2], f2[2], f3[2], f4[2], f[2];
main( )
  {
        // constants for the RKF method
        double const a0 =.25, a1=3./8. , a2=12./13. ;
        double const b10 =.25 ;
        double const b20 =3./32. , b21 = 9./32. ;
        double const b30 =1932./2197. , b31 = -7200./2197. ,
        b32 = 7296./2197. ;
        double const b40 =439./216. , b41 = -8. ,
        b42=3680./513., b43=-845./4104.,

        double const r1 = 25./216. , r2 = 1408./2565. ,
        r3 = 2197./4104. , r4=-1./5. ;
        // end of constants

        // user input
        double spring=0.; // spring constant
        double mass=0.; // mass
        double damp=0.; // damping
        double v_0=0.; // initial condition for velocity
        double x_0=0.; // initial condition for amplitude
        double step=0.; // stepsize
        char filename[80]; // plotfile
        // end user input
```

```
double y[2],y_old[2];
double e_old, e_new;
double f0[2], f1[2], f2[2] , f3[2] , f4[2], f[2];
double result = 0;
double x=0 , x0;
int len=2;
```

//*************************************************************
```
//      here we deal with the user input

        cout << "give the filename for output:";
        cin >> filename;
```

```
//      Now open the file

        ofstream out_file(filename) ;// Open file
                out_file.setf(ios::showpoint);
                // keep decimal point

        cout << "give the step size:";
        cin >> step ;
        cout << "mass and damping";
        cin >> mass >> damp;
        cout << "spring constant";
        cin >> spring;
        cout << "initial position and velocity";
        cin >> x_0 >> v_0 ;
        omega2=spring/mass;
        gam=damp/(mass*2.);
```

//*************************************************************
```
//      input is finished
//      let's do the calculation
```

```
//      use the initial conditions;
        x=0.;
         y[0]=x_0;
         y[1]=v_0;
         y_old[0]=y[0];
         y_old[1]=y[1];

        while(x<20.)
        {
//       here starts the loop
         x0=x;
         deri (x0, &y[0], &f[0], omega2,gam);
          x=x0+a0*step;
```

```
          y[0]=y_old[0]+b10*step*f[0];
          y[1]=y_old[1]+b10*step*f[1];
      deri (x, &y[0], &f1[0], omega2,gam);
        x=x0+a1*step;
          y[0]=y_old[0]+b20*step*f[0]+b21*step*f1[0];
          y[1]=y_old[1]+b20*step*f[1]+b21*step*f1[1];
      deri (x, &y[0], &f2[0], omega2,gam);
        x=x0+a2*step;
          y[0]=y_old[0]+b30*step*f[0]+b31*step*f1[0]+
          b32*step*f2[0];
          y[1]=y_old[1]+b30*step*f[1]+b31*step*f1[1]+
          b32*step*f2[1];
      deri (x, &y[0], &f3[0], omega2,gam);
        x=x0+step;
          y[0]=y_old[0]+b40*step*f[0]+b41*step*f1[0]
          +b42*step*f2[0]+b43*step*f3[0];
          y[1]=y_old[1]+b40*step*f[1]+b41*step*f1[1]
          +b42*step*f2[1]+b43*step*f3[1];
      deri (x, &y[0], &f4[0], omega2,gam);
          y[0]=y_old[0]+step*(r1*f[0]+r2*f2[0]+
          r3*f3[0]+r4*f4[0]);
          y[1]=y_old[1]+step*(r1*f[1]+r2*f2[1]+
          r3*f3[1]+r4*f4[1]);
//        cout << x << " " << y[0] <<" " << y[1] << "\ n";
          out_file << x << " " <<y[1] <<"\ n";
          y_old[0]=y[0];
          y_old[1]=y[1];
          }
//        cout << f[0] << " " <<f[1] <<"\ n";
//        cout << f1[0] << " "<<f1[1] <<"\ n";
          out_file.close();

          }

void deri( double x, double y[ ], double f[ ], double omega2,
double gam)
{
      f[0]=y[1];
      f[1]=-omega2*y[0]-2*gam*y[1];
      return;
}
// compile g++ damp.cxx
```

## rk4_step.cxx

```cpp
// Runge Kutta with adaptive step size control
#include <iostream.h>

void rk4_step( double *y, double *ydiff, double *y_old,
               int nODE, double step, double x0 )
{
        void deri ( int , double , double *, double *);
        // user supplied routine
        // which calculates derivative
        // setup the constants first

        //the coefficients for the steps in x
        double const a2=1./5. , a3=3./10., a4=3./5., a5=1. ,
        a6=7./8. ;

        // coefficents for the y values
        double const b21=1./5. ;
        double const b31=3./40. , b32=9./40. ;
        double const b41=3./10. , b42=-9./10., b43=6.5. ;
        double const b51=-11./54. , b52=5./2. , b53=-70./20. ,
        b54=-35./27.;
        double const b61=1631./55296. , b62=175./312. ,
        b63=575./13824. ;
        double const b64=44275./110592. , b65=253./1771. ;

        // coefficents for y(n-th order)
        double const c1=37./378. , c2=0. , c3=250./621.,
        c4=125./594., c5=0., c6=512./1771. ;

        // coefficents for y(n+1-th order)
        double const cs1=2825./27648. , cs2=0. ,
        cs3=18575./48384., cs4=13525./55296. ,
        cs5=277./14336., cs6 = 1./4. ;

        // the calculated values f
        double f[nODE] , f1[nODE] , f2[nODE] , f3[nODE] ,
        f4[nODE] , f5[nODE];

        // the x value
        double x;
        double yp[nODE];
        int i;

        // here starts the calculation of the RK parameters
           deri (i,x,y,f);
```

```
for(i=0; i<=nODE-1;i++) // 1. step
{
  y[i]=y_old[i]+b21*step*f[i];
}
x=x0+ a2*step;
  deri(i,x,y,f1);
for(i=0; i<nODE-1;i++) //2. step
{
  y[i]=y_old[i]+b31*step*f[i]+b32*step*f1[i];
}
x=x0+ a3*step;
  deri(i,x,y,f2);
for(i=0; i<=nODE-1;i++) //3. step
{
  y[i]=y_old[i]+b41*step*f[i]+b42*step*f1[i]
  +b43*step*f2[i];
}
x=x0+ a4*step;
  deri (i,x,y,f3);
for(i=0; i<=nODE-1;i++) //4. step
{
  y[i]=y_old[i]+b51*step*f[i]+b52*step*f1[i]
  +b53*step*f2[i]+b54*step*f3[i];
}
x=x0+ a5*step;
  deri (i,x,y,f4);
for(i=0; i<=nODE-1;i++) //5. step
{
y[i]=y_old[i]+b61*step*f[i]+b62*step*f1[i]
+b63*step*f2[i]+b64*step*f3[i]+b65*step*f4[i];
 }
x=x0+ a6*step;
  deri (i,x,y,f5);
for(i=0; i<=nODE-1;i++) //6. step
{
  y[i]=y_old[i]+step*(c1*f[i]+c2*f1[i]+c3*f2[i]
  +c4*f3[i]+c5*f4[i]+c6*f5[i]);
  yp[i]=y_old[i]+step*(cs1*f[i]+cs2*f1[i]+cs3*f2[i]
  +cs4*f3[i]+cs5*f4[i]+cs6*f5[i]);
  ydiff[i]=fabs (yp[i]-y[i]);
 }
return;
}
```

### rk4_stepper.cxx

```
// routine rk4_stepper
// adaptive step size Runge Kutta ODE solver
// uses rk4_step
#include <iostream.h>
#include <fstream.h>
#include <math.h>
#include <minmax.h>
#include "TROOT.h"
#include "TApplication.h"
#include "TCanvas.h"
#include "TLine.h"
#include "TPaveLabel.h"
#include "TRandom.h"
#include "TH1.h"
#include "TH2,h"
#include "TH3.h"
#include "TPad.h"

void rk4_stepper(double y_old [ ], int nODE, double xstart,
                 double xmax, double hstep, double eps,
                 int&nxstep)
{
        void rk4_step (double *, double *, double *, int ,
        double ,double);
        double heps; // the product of the step size and the
        chosen error
        double yerrmax=.99; // the max error in a step,
        int i=0;
        double const REDUCE=-.22 // reduce stepsize power
        double esmall ; // the lower limit of precision, if the
        result is smaller than this we increase the step size
        double ydiff[nODE];
        double y[nODE];
        double hnew; // new step size
        double x0;
        double xtemp; // temporary x for rk4_step
        double step_lolim; // lower limit for step size
        double step_hilim; // upper limit for step size

        // here we create a file for storing the calculated
        functions
        ofstream out_file;
        out_file.open ("rk4.dat");
```

```
out_file.setf(ios::showpoint | ios ::scientific |
ios::right);

x0=xstart;
        for(i=0 ; i<=nODE-1 ; i++)
         {
        // store starting values
                out_file << x0 << " "<<Y_old[i]<<"\ n";
          }
esmall=eps/50.;
heps=eps*hstep;

step_lolim=hstep/10.; // the step size should not go
lower than 10 %
step_hilim=hstep*10.; // We don't want larger step size
        for(i=0; i<=nODE-1;i++)
         {
        y[i]=y_old[i];
         }
        for( ; ; )
         {
          yerrmax=.99;
           for( ; ; )
              {
                xtemp=x0+hstep;
                rk4_step(y,ydiff,y_old,nODE, hstep,
                xtemp);
                for(i=0; i<=nODE-1;i++)
                 {
                   yerrmax=max(yerrmax, heps/ydiff[i]);
                 }
                if (yerrmax > 1.) break;
                hstep=.9*hstep*pow (yerrmax, REDUCE);

                // error if step size gets too low
                if (hstep<step_lolim)
                  {
                    cout << "rk4_stepper: lower step
                    limit reacher; try
                    lower starting"
                        << "step size\ n" ;
                    cout << "I will terminate now \ n";
                    exit(0)
                  }
              }
```

```
                                   // go further by one step
                                   // first check if our step size is too
                                   small

                                   if (yerrmax>1./esmall)hstep=hstep*2.;

                                   // set upper limit for step size
                                   if (hstep>step_hilim) {
                                   hstep=step_hilim;
                                   }
                                   for (i=0 ; i<=nODE-1 ; i++)
                                     {
                                        y_old[i]=y[i];
                                   // store data in file rk4.dat
                                        out_file << xtemp << "
                                        "<<y[i]<<"\ n";
                                     }
         //                        x0 += hstep;
                                   x0=xtemp;
                                   nxstep=nxstep+1;
                                   if(x0>xmax)
                                   {
                                   cout << nxstep;
                                     out_file.close(); // close data file
                                     return ;
                                   }
                         }
                       return;

         }
```

## D.3  Random walk in two dimensions

rnd_walk2.cxx

```cpp
// rnd_walk2.cxx
// Random walk in two dimension
// ak 4/30/2002
#include <iostream>
#include <fstream>
#include <math.h>
#include <iomanip>
#include <stdio.h>
#include <stdlib.h>
#include <fcntl.h>

#include "TROOT.h"
#include "TTree.h" // we will use a tree to store our values
#include "TApplication.h"
#include "TFile.h"

int rand(void);
void Srand(unsigned int);

void main(int argc, char ** argv)
{
        //structure for our random walk variables
        struct walk_t { double x; // the position after a step
                        double y; //
                        int nstep; // the step number
                        int jloop; // number of outer loops
                        };
        walk_t walk;

        int i_loop; //inner loop
        int j_loop; //outer loop
        int jloop_max=5000; // the max number of
        different trials

          unsigned int seed = 557 ; // here is the starting value
          or seed

            int loop_max=100; // the maximum number of steps

        double rnd1;
        double rnd2;
        double y_temp; // temporary variable for y
```

```
TROOT root ("hello", "computational physics");

// initialize root
TApplication theApp("App", &argc, argv);

//open output file
TFile *out_file = new TFile("rnd_walk557_2.root",
"RECREATE","example of random random walk");
// create root file

// Declare tree
TTree *ran_walk = new TTree("ran_walk","tree with
random walk variables");
ran_walk->Branch("walk",&walk.x, "x/D:y/D:nstep/
I:jloop/I");

// set seed value
srand(seed);

// the outer loop, trying the walk jloop times
for (j_loop=0;j_loop < jloop_max ; j_loop= j_loop+1)
{
        walk.x=0.;
        walk.y=0.;
        walk.nstep=0;
        walk.jloop=j_loop+1;

for(i_loop=0; i_loop < loop_max ;i_loop= i_loop+1)
{
 // here we get the step
        rnd1=double(rand())/double(RAND_MAX);
         rnd1=2*rnd1-1.;
          walk.x=walk.x+rnd1;

          if(rnd1*rnd1>1.) rnd1=1.; //safety for
          square root
          Y_temp=sqrt(1.-rnd1*rnd1);

        rnd2=double(rand())/double(RAND_MAX);
         if((rnd2-.5)<0.)
          {
          walk.y=walk.y-y_temp;
          }
          else
          {
          walk.y=walk.y+y_temp;
          }
```

```
      walk.nstep=walk.nstep+1;
      // fill the tree
       ran_walk->Fill();
   }
  }
    out_file->write();
}
```

## D.4 Acceptance and rejection method with sin(*x*) distribution

rnd_accept.cxx

```
// rnd_invert.cxx
// Random number using the acceptance / inversion method
// this simple program uses the sin function as the probability
// distribution
// ak4/30/2002
#include <iostream>
#include <fstream>
#include <math.h>
#include <stdio.h>
#include <stdlib.h>
#include <fcntl.h>

#include "TROOT.h"
#include "TTree.h" // we will use a tree to store our values
#include "TApplication.h"
#include "TFile.h"
#include "TF1.h"

int rand(void);
void srand(unsigned int);

void main(int argc, char ** argv)
{
        double x_low=0.; // lower limit of our distribution
        double x_high=180.; // the upper limit
        double deg_rad=3.14159265/180.; //converts degrees in
        rads

        //structure for our distribution variables
        structure dist_t { doubt x_throw; // the thrown value
                           double x_acc; // the accepted value
                           double y_throw;
                           double y_acc;
                           };
        dist_t dist;

        int i_loop; //inner loop

          unsigned int seed = 68910 ; // here is the starting
          value or seed
          int loop_max=1000000; // the maximum
          number of steps
```

```
    double rnd;

TROOT root ("hello", "computational physics");

// initialize root
    TApplication theApp(*App*, &argc, argv);

//open output file
TFile *out_fill = new TFile("rnd_acc.root",
"RECREATE", "A distribution following a sine");
 // create root file

// Declare tree
TTree *dist_tree = new TTree("dist_tree","tree
with rejection");

 dist_tree->Branch("dist", &dist.x_throw,
 "x_throw/D:x_acc/ D:y_throw/D:y_acc");

// set seed value
srand(seed);

for(i_loop=0;i_loop < loop_max= i_loop=i_loop+1)
{
// step 1: throw x between 0 and 180
dist.x_throw=x_low+double(rand())/double(RAND_MAX)*
 (x_high-x_low)*deg_rad;

//step 2: create a random variable in Y between 0 and 1
        dist.y_throw=1.*double(rand())
        /double(RAND_MAX); // from 0,1
//step 3: Check it f(x)>y and if true accept
        if(sin(dist.x_throw)>dist,y_throw)
         {
           dist.x_acc=dist.x_throw/deg_rad;
            dist.y_acc=dist.y_throw;
        else
// these are the rejected ones.
          {
          dist.x_acc=-999;
          dist.y_acc=-999;
          }
          dist_three->Fill();
    }
        out_file->Write();

}
```

# Index